NIST Technical Note 1481

Ignition of Cellulosic Fuels by Heated and Radiative Surfaces

William M. Pitts
Building and Fire Research Laboratory

I0482663

March 2007

U.S. Department of Commerce
Carlos M. Gutierrez, Secretary

Technology Administration
Robert Cresanti, Under Secretary of Commerce for Technology

National Institute of Standards and Technology
William Jeffrey, Director

Abstract

Experiments designed to characterize the ignition behavior of typical outdoor fuels by heated mufflers and catalytic converters found on outdoor power equipment are described. Ignition by direct contact with a heated surface and by radiation from a heated surface were considered. For the first scenario the fuel was placed in contact with an electrically heated surface, and the times required for ignition measured as a function of surface temperature. The effects of a wind on heated plate ignition were studied by passing air flows of 1.1 m/s and 2.5 m/s over the fuel bed. Fuels tested included shredded newsprint, four types of grass, pine needles, a grass/pine needle mixture and three types of dry leaves. Both glowing combustion and flaming were observed. The transition to flaming required glowing combustion to be present. Ignition times generally increased with decreasing surface temperature until a temperature was reached where ignition was no longer observed. Ignition times for given applied wind and surface temperature conditions and transition to flaming behavior were fuel dependent. An applied wind generally reduced the time required for ignition, with the reduction being greater for the higher velocity. A commercially available cone calorimeter operated in the non piloted ignition mode was used to impose known radiative heat fluxes on the fuel surface, and ignition times were measured as a function of applied heat flux. Shredded newsprint, two types of grass, pine needles, and non fire-retarded flexible polyurethane foam were studied. Ignition times for radiative heating increased with decreasing applied heat flux until levels were reached where ignition did not occur. The polyurethane foam had a distinctly different behavior than the cellulosic fuels. Comparison of the heated plate and radiative heating experiments suggests that they can be related using the black body temperature corresponding to the applied radiative heat flux.

Table of Contents

List of Tables

List of Figures

Executive Summary

Recently implemented and proposed regulations for outdoor power equipment require reductions in hydrocarbon and nitrogen oxides to levels that require the use of catalytic converters. Concerns have been raised that catalytic converters can result in increased temperatures for exhaust gases and surfaces that could increase the risk of fires. In response to this concern, The International Consortium for Fire Safety, Health, and the Environment (ICFSHE), with funding provided by the Outdoor Power Equipment Institute, awarded a contract to the SP Swedish National Testing and Research Institute (SP) for a "Scientific Evaluation of the Risk Associated with Heightened Environmental Requirements on Outdoor Power Equipment." As part of this study ICFSHE requested that the Building and Fire Research Laboratory of the National Institute of Standards and Technology (BFRL/NIST) provide experimental support to SP. A work statement was adopted that involved characterizing the ignition of typical outdoor fuels by ignition sources representative of those expected for outdoor power equipment exhaust systems.

This report summarizes the findings of the BFRL/NIST investigation. One series of experiments was designed to simulate the ignition behaviors of fuels that come into direct contact with a heated surface. A series of experiments was run in which porous fuel beds were brought into contact with a 10.2 cm × 10.2 cm heated copper plate held at a constant temperature. The plate faced upward. A constant mass of the material to be tested was placed in a stainless steel wire screen cage to a depth of 2.5 cm, placed over the heated plate, and then exposed to the surface by removing a metal shutter from the bottom of the cage. The primary experimental measurements were whether or not glowing and/or flaming developed and, if so, the heating times required. Observations were made visually and by video recording of the experiments. Secondary observations included smoke behaviors, locations of initial burning, and the appearance of the fuel bed following an experiment. The effect of a wind on ignition behavior was investigated by running experiments with no wind and in the presence of low (1.0 m/s) and high winds (2.5 m/s). Heated plate temperatures were varied from temperatures in excess of 500 °C where ignition periods were short (a few seconds) down to temperatures where ignition of the fuels was not observed after many minutes.

A second series of experiments was designed to characterize ignition of porous fuel beds by radiative heating. Such heating occurs when fuel is located near, but not in direct contact with, a heated surface. The experiments were made with a cone calorimeter, a standard instrument designed primarily for heat release rate measurements with an applied radiative heat flux. Cone calorimeter measurements usually utilize an electrical spark as an ignition source, but these experiments were non piloted, and the spark was not used. Exposed fuel bed surfaces had the same dimensions as for the heated plate experiments, but samples were placed in sample holders with solid walls and were irradiated from above by a radiant source that was shielded from the fuel by a cooled shutter until the start of an experiment. Glowing and flaming ignition times were determined for a range of applied heat fluxes, from levels sufficient to induce ignition within a few seconds down to levels where ignition was not observed after many minutes. The only air flow was that around the sample induced by the experimental system. Useful secondary measurements included the heat release rate and mass for the sample holder and fuel as functions of time.

Heated plate ignition experiments were done for shredded newsprint, four grasses (two samples of tall fescue collected from the same location at NIST in May and August, cheat grass from California, and a sample of fine grass from Florida provided by a manufacturer of outdoor power equipment that was intended to simulate debris from the top of a mower housing), pine needles, and a mixture of May tall fescue, and pine needles. A limited number of tests were also run for boxwood, American elm, and pin oak leaves. Cone calorimeter tests were done for shredded newsprint, May tall fescue, cheat grass and pine needles. Samples of a non fire-retarded flexible polyurethane foam, representing a common plastic, were also tested in the cone calorimeter.

Some general observations from the heated plate experiments include the release of smoke, most likely a combination of condensed water and organic pyrolyzate as fuels are heated. The smoke distribution for experiments without a wind indicated that circular buoyant plumes were formed within the porous fuel beds by the flows rising from the heated plate and oxidizing fuel surfaces. With a wind applied to the fuel bed, smoke escaped along a band extending across the fuel bed perpendicular to the wind direction. This smoke distribution resulted from the interaction of the wind entering the fuel bed, which was slowed by interaction with the fuel, and the buoyancy driven flow. Qualitative observation indicated that the amount of pyrolysis occurring within a fuel bed was related to the amount of smoke observed. Smoke levels were generally heavy immediately prior to glowing or flaming ignition.

For heated plate temperatures around 550 °C all of the fuels ignited within a few seconds. As the plate temperature was reduced, the ignition times at first slowly increased until temperatures reached about 450 °C. Further reductions in temperature resulted in ignition times that increased more rapidly as the plate temperature was reduced. The dependence on temperature was shown to be captured well by fitting with exponential curves. Eventually, as the plate temperature was lowered, a point was reached for which a given type of fuel bed no longer ignited.

The ignition described above includes both glowing combustion and flaming. There were conditions for which glowing was observed but flaming was not and fewer examples of flaming without observed glowing. Even though glowing was not observed in all experiments in which flames were seen, in most experiments it appeared earlier. Video recordings of the transition to flaming showed that flames ignited within regions surrounded by intense glowing. A blue flame would suddenly appear and then spread rapidly before yellow flames appeared. These observations suggest that the transition to flaming requires the presence of high temperature glowing areas to supply the ignition energy and premixed volumes of gaseous fuel pyrolyzate and air having concentrations within the flammability limits for the mixture.

The general heated plate ignition temperature dependence discussed above was observed for the three wind conditions tested, but ignition times and fuel dependent transition to flaming behaviors depended strongly on wind condition. When a low wind was applied the ignition times were reduced from those for the same fuel without a wind, assuming ignition occurred for both conditions. Similar reductions in ignition times were observed when results for low and high winds were compared. For the lowest plate temperatures, the reductions in ignition times on going from no wind to high wind were on the order of a factor of two. Transition from glowing combustion to flaming was observed whenever glowing developed for shredded newsprint and pine needles. With a few isolated exceptions, none of the grass samples that developed glowing in the absence of wind transitioned to flaming, while the August tall fescue, cheat grass, and fine Florida grass did transition to flaming when low and high winds were present. For most cases, glowing May tall fescue failed to transition to flaming with wind present. Note that the difference between the May tall fescue and August tall fescue implies seasonal differences for grass samples taken from the same location. The May tall fescue/pine needle mixture displayed flaming behaviors intermediate between those for the individual fuels, transitioning from glowing to flaming in over half of the experiments without wind and in all cases when wind was present.

Ignition times for the individual fuels displayed similar dependencies on heated plate temperature and wind as described above. By fitting the combined results for each wind condition to an exponential, it was possible to demonstrate that systematic effects of fuel were present that introduced ignition time variations that were larger than those observed for a given individual fuel. For a specified heated plate temperature and wind condition, the longest ignition times were found for the shredded newsprint and pine needles, while the grasses tended to have the shortest ignition times. The May tall fescue/pine needle mixture gave intermediate ignition times.

Minimum heated plate temperatures required for a given fuel and wind condition can be estimated from the experimental measurements. Values ranged from 290 °C to 380 °C and had complicated dependencies on fuel type and wind condition. These dependencies are described in detail in the full manuscript. In general, when a wind was not present the grasses tended to have higher minimum heated plate ignition temperatures than the shredded paper and pine needles. When wind was introduced the minimum temperatures for the grasses decreased while those for the shredded paper and pine needles increased such that the lowest minimum temperatures with wind tended to be those for the grasses. Inspection of the grass fuel beds following experiments at temperatures lower than required for ignition with no wind provided insight into the reason for this behavior. It was found that non glowing smoldering took place in the absence of a wind down to temperatures on the order of 300 °C. Presumably, when a wind was applied more oxygen was available at the fuel surface and sufficient heat was generated for the non glowing combustion to transition to glowing. Similar non glowing smoldering was not observed for shredded paper and pine needles. It is likely that for these fuels the applied wind cooled the samples and resulted in the higher minimum ignition temperatures. The May tall fescue/pine needle mixture without wind yielded the lowest observed minimum ignition temperature of 290 °C. This suggests that the May tall fescue component of the mixture was able to begin non glowing smoldering at this temperature and that the heat generated was sufficient to ignite glowing combustion on the pine needle fraction. Transition to flaming was observed in this particular experiment.

Limited experiments with a heated plate temperature of 380 °C indicated that leaves are likely to have similar ignition behaviors to those observed for the cellulosic fuels investigated more fully.

The cone calorimeter radiative ignition (glowing and flaming) experiments were run without a lateral air flow. Applied heat fluxes were varied from a high of 50 kW/m^2 down to values where ignition was not observed. Plots of ignition time versus applied heat flux had similar appearances to the plots of ignition time versus heated plate temperature discussed above. For the highest heat fluxes ignition times were on the order of a few seconds. As the applied heat flux was reduced to lower values, the ignition times increased slowly at first. Eventually, with further reductions, applied heat fluxes reached values where ignition times began to increase rapidly at rates that increased with decreasing flux. Ignition times of hundreds of seconds were observed before further reductions in applied heat flux resulted in no ignition. As for the heated plate experiments, the dependencies of ignition times on applied heat flux were well described by exponential curve fits.

Results for three of the fuels, shredded paper, May tall fescue, and cheat grass, fell on very similar curves, with the rapid increases in ignition beginning around 20 kW/m^2 and minimum applied heat fluxes required for ignition of approximately 10 kW/m^2. The ignition time for the pine needles was slightly different, with the rapid rise starting around 30 kW/m^2 and a slower rate of increase with decreasing heat flux than found for the three other fuels. A minimum applied heat flux for ignition was not determined for the pine needles.

Even though the ignition times as a function of applied heat flux were similar for the four fuels, transition to flaming behaviors were very different. Flaming was observed for all of the fuels with applied heat fluxes of 40 kW/m^2 and higher, and maximum observed heat release rates were similar. However, as the applied heat flux was reduced below 40 kW/m^2 only glowing combustion developed for the two grasses, while the shredded newsprint and pine needles continued to transition to flaming. These ignition behaviors were observed over applied heat fluxes in the 10 kW/m^2 to 40 kW/m^2 range. Observed time behaviors for heat release rate and mass loss were consistent with these observations. The highest heat release rates and mass loss rates were observed during flaming, while the values were much reduced during smoldering. The measurements for the glowing grass indicated that there were two distinct types of smoldering having different effective heats of combustion. There seemed to be an initial oxidation that converted a part of the grass to a high energy char, which then smoldered slowly with much slower mass loss but a comparable heat release rate.

The cone calorimeter results are in good agreement with the heated plate results with regard to flaming behavior for the grasses. They indicate that with no wind present it is very difficult to generate flaming, consistent with the absence of flaming in the heated plate experiments. The shredded newsprint and pine needles transitioned to flaming in all cases where glowing developed, again consistent with the heated plate experiments. Measurements of ignition times, heat release rates, and mass loss were similar for the May tall fescue and cheat grass. Recall that these two grasses had different behaviors with regard to transition to flaming in the presence of a wind. The similarity of the results suggests that the cone calorimeter is not able to distinguish fuels that have different flaming behaviors in the presence of wind.

Radiative ignition experiments were performed for polyurethane foam. The ignition time behavior was very different from those observed for the cellulosic fuels, and the minimum ignition applied heat flux was 23 kW/m^2. Heat release rate curves had two peaks corresponding to initial burning of the foam that generated a liquid component, which then burned as a pool fire. At the lowest applied heat fluxes the foam formed the liquid without oxidation, and if flaming ignited, it was only for the liquid fraction.

In order to determine if the response of a fuel to a radiative heat flux could be related to its response on a heated plate, black body temperatures corresponding to given applied heat fluxes were calculated. Comparisons of heated plate temperatures and the corresponding black body temperatures from the cone calorimeter results indicated that values for minimum ignition temperature and the temperature where the sharp increases in ignition times began were slightly higher for the black body temperatures, but were close enough to justify the approach. This is particularly true when it is noted that the application of heat to the bottom of the fuel beds and open walls of the fuel holder for the heated plate experiments should result in slightly lower ignition temperatures than the radiative heating experiments where heat is applied to the top of fuel beds enclosed on the sides.

A review of the literature identified very few experiments that could be compared directly to the current findings. Limited numbers of measurements of ignition times versus surface temperature were consistent for similar fuel beds. A review of experiments that identified minimum ignition temperatures yielded a wide range of values that were of little use for comparison. No experiments were identified that investigated non piloted radiative ignition of these types of fuels.

The experimental results are subject to uncertainties associated with such parameters as temperature measurement, identification of ignition time, and stochastic variations of fuel behavior and packing. Due to the limited number of measurements possible, it was not feasible to determine probability curves for such variables as ignition temperature as a function of heated plate temperature or applied heat flux. In spite of these uncertainties the results demonstrate that smoldering and flaming ignition of cellulosic fuel beds exposed to conditions chosen to be representative of those generated on heated exhausts of outdoor power equipment vary with surface temperature or applied heat flux, fuel type, wind, and season. In addition to these parameters, there are a number of others that would be expected to affect ignition behavior but were not varied in this study. These include fuel moisture, fuel bed exposed surface area, fuel bed thickness, fuel bed porosity, spatial orientation of the heat source and the fuel bed, and the accessibility of the interior of a fuel bed to an applied wind.

Keeping in mind the experimental uncertainties and additional parameters that could affect ignition behavior, it is possible to provide some general guidance with regard to expected ignition behaviors of cellulosic fuels when subjected to contact with a heated surface or exposure to thermal radiation. Curve fits to exponentials based on data for multiple fuels have been made that provide reasonable approximations for the observed variations in ignition times with heated plate temperature and applied heat flux. For the heated surface, these fits are available for three wind conditions. While clearly not representative of worst case conditions, these curves should be useful for assessing the likelihood of ignition and the time required for a generic cellulosic fuel subjected to conditions similar to those investigated here.

It is common to introduce safety factors when making engineering estimates having safety implications. For the ignition of cellulosic fuels by contact with a heated surface the following assumptions are appropriate: 1) ignition is possible for temperatures of 290 °C and higher, 2) any ignition that takes place will result in flaming, and 3) ignition times estimated from the exponential curve fits should be divided by a factor of two. The available experimental data cover a range of winds from 0 m/s to 2.5 m/s, and the fits can be used to estimate ignition times over this range. Caution should be exercised when extending the estimates to higher wind cases.

An exponential curve fit of ignition times versus applied heat flux has been determined from the radiative heating experiments. Similar assumptions to those above would allow this curve to be utilized for estimating ignition times for cellulosic fuel beds exposed to a given level of heat flux. However, based on the similar ignition time dependencies observed between heated plate temperatures and black body temperatures derived from the applied heat fluxes, it is recommended that black body temperatures based on the applied heat fluxes be used in conjunction with the exponential curves derived from the heated plate experiments to estimate ignition times for radiative heating. This has the advantage of providing estimates for ignition by radiative heating in the presence of a wind.

Suggestions are provided for improvements to the experimental system and for additional experiments.

1. Introduction

Concerns about the levels of hydrocarbons and nitrogen oxides released from the exhausts of outdoor power equipment (OPE) led the California Air Resources Board to approve regulations in September, 2003 requiring reductions in these species to levels which require the use of exhaust catalytic converters. Having received a required US Environmental Protections Agency (EPA) waiver, these regulations went into effect in 2007. The EPA is currently considering extending a similar regulation to the entire country. [1]

Exhaust catalysts operate by lowering the temperatures required to oxidize pollutant species. Since oxidation reactions are exothermic, heat is generated, and there is the potential for exhaust gas temperatures to increase when a catalyst is present. The use of a catalyst may raise the temperatures of the gases exhausted by the engine as well as the temperatures of exposed surfaces that form the exhaust system.

It is known that the temperatures generated by existing exhaust systems on OPE are high enough to be potential fire ignition sources for typical organic fuels. Several groups, including the Outdoor Power Equipment Institute (OPEI) and the National Association of State Fire Marshals (NASFM), questioned whether higher temperatures associated with the use of catalytic converters on OPE might result in an increased risk of unwanted fire.

In response to these concerns and at the request of NASFM, The International Consortium for Fire Safety, Health, and the Environment (ICFSHE), with funding provided by OPEI, awarded a contract to the SP Swedish National Testing and Research Institute (now renamed SP Technical Research Institute of Sweden, referred to as "SP") for a "Scientific Evaluation of the Risk Associated with Heightened Environmental Requirements on Outdoor Power Equipment." Phase I of this investigation included a literature review and development of an experimental research plan for studying this problem. The findings of the Phase I study are available. [2]

ICFSHE requested that the Building and Fire Research Laboratory of the National Institute of Standards and Technology (BFRL/NIST) provide experimental support to SP as part of the Phase II effort. A work statement entitled "Ignition of Fuels by Conductive, Radiative, and Conductive Heating" was prepared and submitted to ICFSHE and SP for consideration. The work statement proposed to experimentally characterize the ignition behaviors of a variety of natural and man made organic fuels when exposed to heat sources believed to be representative of those generated by the exhausts of OPE. This work statement was accepted and research funding was provided to BFRL/NIST by ICFSHE with technical management and oversight provided by SP.

The original work statement included ignition studies using three types of heat sources—heated plate (conductive), heated flow (convective), and radiative. A variety of materials were to be investigated including a grass, pine needles, leaves, sawdust, oil, and gasoline. Tests were proposed to characterize the effects of spatial orientation, an imposed wind, and an uneven heat distribution on the heated plate.

1

Initial testing showed the ignition of grasses was dependent on the type of grass and the presence or absence of a wind played a dominant role in ignition behavior. Based on these observations and with the agreement of SP, the test matrix was modified to focus more attention on the effects of grass type and wind condition. It was decided that characterizing ignition by a heated plate and by radiation was more important than characterizing ignition due to a convective heat flow, and this heat source was dropped from the investigation. Ultimately, ignitions studies on a variable-temperature heated plate were carried out for shredded newsprint, four different grass samples, pine needles, and a grass/pine needle mixture. Three wind conditions—no wind and two imposed wind velocities of 1.1 m/s and 2.5 m/s—were used. A limited number of tests were done with three types of leaves. Radiative ignition studies were performed as a function of applied heat flux for shredded newsprint, two types of grass, pine needles, and non fire-retarded flexible polyurethane foam.

The following section, Section 2, provides an overview of relevant past studies and introduces some concepts useful for the later discussion. The experimental systems and procedures are discussed in Section 3. Section 4 presents the experimental findings for the investigation. The final section, Section 5, summarizes the findings along with the implications for possible ignition by heated plates or radiative surfaces.

2. Important Concepts and Literature Review

The focus of this experimental program is the characterization of ignition for various fuels by heat sources deemed to be representative of those expected for exhaust systems on OPE such as lawn mowers. Two potential scenarios are emphasized. In the first it is assumed the fuel comes into direct contact with a heated surface. Since exhaust temperatures can vary over a wide range, the response of the fuel will be investigated as a function of surface temperature. In the second scenario the fuel is located close to, but sufficiently far from, a heated surface that heating of the fuel is dominated by radiative heat transfer. In this case the important experimental parameter is the radiative heat flux (kW/m^2) impinging on the fuel surface. Since the heat flux depends on the radiating surface temperature to the fourth power, the fuel response to a range of heat fluxes is considered.

Since by definition OPE is operated outdoors, ignition of natural fuels such as grasses, pine needles, and leaves are considered to be particularly important. It is also possible that heated exhaust surfaces will come in contact with a variety of man-made fuels during storage. As a result there is a nearly infinite number of potential fuels that could be considered relevant to this problem. For this study the ignition behaviors of shredded newsprint, four types of grass (May tall fescue, August tall fescue, cheat, and fine Florida), pine needles, various types of leaves, a May tall fescue/pine needle mixture, and polyurethane foam were investigated. While chosen to be representative of possible fuels, care should be exercised in extrapolating the findings to additional fuels.

The focus of this investigation is ultimately fire safety. From this standpoint, "flaming combustion" in which gasified fuel molecules undergo a self sustaining reaction with oxygen in the air to release heat and light is considered to represent the most dangerous type of combustion.

The natural fuels chosen for study as well as the shredded newsprint contain high concentrations of cellulose along with other natural polymers such as hemicellulose, and lignin. Materials containing these polymers are known to be susceptible to a type of surface combustion known as smoldering. In his review, Ohlemiller defines smoldering as "a slow, low-temperature, flameless form of combustion, sustained by heat evolved when oxygen directly attacks the surface of a condensed-phase fuel." [3]

Smoldering usually involves both endothermic and exothermic processes. Moisture removal from the fuel and gasification of low molecular weight fuel molecules (known as pyrolysis) generally require energy input. As described by Ohlemiller, fuel pyrolysis often leads to the formation of a poorly characterized carbon-enriched material referred to as char. Oxidative surface reactions with char are exothermic and provide the heat necessary to sustain smoldering. [3] Note that fuel pyrolysis is possible without oxidation. Local surface oxidation can also occur near a heated surface without the development of self-sustaining smoldering.

Once initiated, in most cases the smoldering reaction rate is limited by the amount of oxygen available at the fuel surface. Therefore smoldering behavior is generally sensitive to parameters that affect the amount of oxygen reaching the surface. An example of such a parameter is the porosity of a smoldering fuel bed. The more open the fuel bed, the more easily oxygen will be able reach a given fuel surface and support smoldering. The presence of an imposed or natural (buoyancy-driven) flow strongly influences the amount of oxygen reaching a smoldering fuel surface. Everyone is familiar with the increase in intensity that occurs when one blows on a piece of paper or wood with visible smoldering. The fuel surface-to-volume ratio is also an important parameter controlling smoldering behavior with high ratios favoring smoldering.

Most studies of smoldering consider such properties as the spread rate in a fuel bed. The study of Palmer, who investigated smoldering spread through a variety of materials and configurations, is a classic example. [4] Studies that consider ignition of smoldering are much rarer. In his comprehensive review on ignition, Babrauskas includes a section on smoldering in general and smoldering ignition in particular. [5] Many of the studies he summarizes involve determination of the minimum energy necessary to ignite smoldering. Of direct relevance to this study, Ohlemiller showed that the lowest temperature of a heated surface capable of igniting smoldering in a bed of cellulosic insulation varied with the size of the heater. [6,7] For a 0.48 cm diameter wire a minimum temperature of 380 °C to 385 °C was required, while the limit dropped to 255 °C to 260 °C for a 200 cm × 200 cm plate. Times required for ignition with these low temperatures were on the order of hours. Babraukas also discusses a limited number of studies in which the time to ignition was determined as a function of applied heat flux. Materials studied included cotton fabrics and polyurethane foam. [5]

A number of parameters have been identified as important in controlling smoldering behavior in porous fuel beds such as those considered in this study. [3,5] The fuel must be capable of forming a rigid char during pyrolysis. Char formation varies widely in solid fuels. Charring and char oxidation rates in cellulosic fuels are known to be sensitive to the concentration of various inorganic salts, metals, and metal hydroxides. The amount of moisture absorbed in the fuel is a very important parameter. Physical properties of the fuel beds such as size, thickness, and porosity also play roles.

Babrauskas notes that there are two smoldering modes that are characterized by whether or not oxidizing surface temperatures are high enough to emit visible light. [5] The type of smoldering in which light is emitted is often referred to as "glowing combustion." Drysdale includes a short discussion of glowing combustion as part of a discussion of smoldering in his book. [8] In the current work smoldering in which there is no visible light produced will be referred to as "non glowing combustion." It appears that little previous research has considered the conditions necessary for the development of glowing combustion or its general behavior.

It is well known in the fire community that under certain conditions smoldering will spontaneously transition to flaming combustion. [5,8] While the general requirements for ignition in the gas phase are known, there is little practical understanding of the transition to flaming from smoldering solids. At a minimum, it is necessary that a solid fuel pyrolyze with a sufficient rate to generate a premixed fuel/air mixture that lies between the lean and rich flammability limits. Such a mixture can ignite in two well defined ways—piloted and non-piloted. In "piloted ignition" a source of heat and free radicals, such as a flame or spark, are used to initiate the combustion reactions, which then become self sustaining. In non-piloted ignition the flammable fuel/air mixture must be heated to a point where free radicals are generated within the mixture, and rapid chain branching chemical reactions lead to sustained combustion. Typically, higher temperatures must be present for non piloted ignition than for piloted ignition.

In the current work a cone calorimeter will be used for the radiative ignition studies. Babrauskas and Parker have described the use of the instrument for this purpose. [9] Most ignition experiments that have been done in the cone utilized an electric discharge as a pilot for the flame. It is possible to operate the system without a spark in order to study non piloted ignition.

A primary focus of this work is the ignition behavior of cellulosic fuels. Babrauskas included a review of ignition of forest materials, vegetation, and hay in his comprehensive monograph on ignition. [10] His general comments concerning such measurements are relevant. He points out that there are numerous measurements of ignition temperatures in the literature, but that the results vary widely. Potential sources of these variations include fuel composition and moisture content, but Babrauskas argues that variations due to such factors should be much less than observed, and he attributes much of the variability to experimental protocols and operator inexperience.

Many of the ignition studies cited by Babrauskas involve determining a lower temperature limit for combustion. A variety of heating approaches were employed with the most prevalent being immersion in air held at a constant temperature in an oven. Minimum ignition temperatures reported for a variety of natural fuels ranged from 166 °C to as high as 675 °C. There is no indication of the times required for ignition or the type of combustion observed.

In a number of studies relatively large heated bodies were deposited on various types of porous natural fuel beds. Ignition temperature tended to be somewhat higher than reported for other ignition sources, with minimum values ranging between 450 °C and 650 °C.

More recently, several studies of radiant ignition of forest fuels have appeared. Most of these tests have utilized either the ISO 5657 [11] or cone calorimeter [12,13,14] tests. ISO 5657 and the cone calorimeter are similar tests, but differ in the sample size and configuration and in the type of igniter used in piloted ignition studies. Most of these tests were designed to measure the flammability of natural fuels and usually utilized piloted ignition. Some findings are summarized by Babruaskas, but it is difficult to draw conclusions relevant to the current investigation. [10]

In a book on the investigation of wildfires, Ford includes a useful summary of selected studies on wildland fuel ignition. [15] He discusses similar large variations in minimum ignition temperatures to those cited by Babrauskas, but notes wildland physicists have settled on a temperature of 320 °C as an arbitrary average. Ford summarizes the results of an extensive investigation of ignition by heated surfaces carried out at the Oregon State Engineering Research Station in 1949. Lower ignition temperatures for a wide range of woods and papers fell in a range of 220 °C to 300 °C.

There is an extensive literature available concerning the pyrolysis and burning behavior of cellulose and cellulosic fuels. The observed behaviors are quite complex and a detailed discussion is not appropriate for the purposes of this investigation. However, it is possible to provide some guidance for the general effects of fuel composition on burning behavior. The recent review of Plucinski is useful for this purpose. [16] Quoting this author, "Plant tissue is approximately 50 % carbon, 44 % oxygen and 5 % hydrogen by mass and varies between 41-53 % cellulose, 16-33 % hemicellulose, and 16 – 33 % lignin. Cellulose tends to burn by flaming combustion and lignin by glowing combustion."

In the following, several investigations are briefly summarized because they used experimental approaches similar to those of this study or considered similar fuels.

In a study primarily concerned with the temperatures generated by automobile exhaust systems, Harrison reported ignition times for a dry annual grass and dry pine needles as functions of the temperature of a heated barbeque charcoal igniter placed on top of piles of the fuels. [17] For both fuels the ignition times increased with decreasing temperature, with the times becoming longer as the temperature was lowered. For the grass the minimum ignition temperature was 400 °C and for the pine needles the corresponding temperature was 350 °C. Both fuels required about 4 min for ignition at these minimum temperatures. Ignitions at lower temperatures were not observed in experiments lasting six minutes. A 0.9 m/s (2 mph) wind had a noticeable effect on grass ignition time, reducing the ignition time for a 400 °C surface to 1.3 min.

In a particularly relevant investigation, Kaminski simulated the ignition of punky (decayed, crumbly, and dry) wood, cheat grass, saw dust, and tree moss by a chain saw muffler by attaching a wire heater to an actual muffler. [18] It was found that punky wood would smoke and glow at temperatures as low as 270 °C. The presence of a low wind reduced the time required for ignition and increased the intensity of the glowing combustion. Glowing for cheat grass was observed at 330 °C, but sawdust only browned under these conditions. The tree moss was reported to have a behavior similar to the punky wood.

In a series of reports Stockstad reported on non piloted and piloted ignition studies of rotten wood [19], cheat grass [20], and pine needles [21] in which minimum heat flux intensities for ignition, time to ignition, and temperatures at times of ignition were given. Small individual pieces of the samples were heated by immersion in a furnace. A variety of rotten woods were tested. Minimum temperatures for ignition were on the order of 270 °C to 300 °C. Times required for glowing combustion were on the order of 1 min. For cheat grass the minimum temperature for which non piloted ignition was observed was 450 °C with ignition times on the order of 20 s. Ignition of pine needles occurred for temperature as low as 365 °C, with glowing appearing after roughly 60 s.

Recently, Manzello et al. have investigated the ignition of shredded paper, pine needle, grass, and hardwood mulch fuel beds by glowing and flaming firebrands deposited on top of the beds. [22,23] Experiments were done with single and groups of four firebrands of two sizes (diameters of 25 mm and 50 mm). Imposed air flows of 0.5 m/s and 1.0 m/s were used. The effect of moisture level was tested by using 0 % and 11 % fuel moisture levels. Ignition by glowing brands is most relevant to the current study, and these findings can be summarized as follows. The only fuel that was readily ignited by a single firebrand was the shredded paper. Only glowing combustion was observed. When 4 brands were used the pine needle fuel beds ignited only with the 50 mm-diameter brands in the presence of the 1 m/s wind. In this case smoldering ignition with transition to flaming was observed for both fuel moistures. The larger firebrands induced smoldering in the dried hardwood mulch with both airflows, with the smoldering transitioning to flaming for the higher air flow. Grass only ignited when dry and exposed to four brands and the higher air flow. Combustion began as smoldering and transitioned to flaming.

White and coworkers have described on an effort to develop an approach for flammability measurements of vegetation using a cone calorimeter. [24,25,26] Measurements have been typically made with applied heat fluxes of 25 kW/m^2 or 50 kW/m^2 and piloted ignition. Changes in flammability due to seasonal variations for several species of ornamental vegetation [25] and for cheat grass grown in the presence of varying CO_2 concentrations [26] have been reported.

3. Experimental Systems and Procedures

3.1. Ignition by Thermal Conduction from a Heated Surface

Ignition of cellulosic fuels as the result of thermal conduction from a heated surface was investigated by placing loosely packed samples of the material to be tested in contact with a uniformly heated surface. The surface was oriented facing upward. The response of the sample to heating was observed while maintaining a constant surface temperature. The dependence of a given fuel response on surface temperature was determined by repeating the experiment for a range of surface temperatures. Experiments were done with and without a lateral air flow applied to the fuel bed.

3.1.1. Experimental System

3.1.1.1. Heated Plate

The heated plate was fabricated by inserting four Watlow[*] 300 W cartridge heaters (Model number E4A30) into a 10.2 cm × 10.2 cm square plate having a thickness of 1.9 cm. Machinable copper was used for the plate due to its high thermal conductivity, which minimized temperature gradients within the plate. Each heater was 10.2 cm long and had a diameter of 0.63 cm. The four heaters were inserted into four parallel 0.64 cm-diameter holes, drilled horizontally through the plate at the vertical center, and positioned 0.95 cm, 3.5 cm, 6.7 cm, and 9.2 cm from the perpendicular edge of the plate. The heaters were equipped with male National Pipe Thread (NPT) fittings and were inserted firmly into machined female NPT fittings in the copper plate. A schematic of the plate showing planar and side views is shown in Figure 1.

Two power supplies were utilized to provide electric current for the heaters. One of the sources was a Kepco Model ATE 100-2.5M capable of supplying up to 2.5 A at a maximum of 100 V DC. The second power supply was a Kepco Model ATE 150-7M that could provide up to 7 A with a maximum voltage of 150 V DC. One of the center cartridge heaters was connected to the smaller power supply, while the three remaining heaters were wired in parallel to the larger supply. Assuming that the four cartridge heaters are operated at the same current, this arrangement could provide a maximum current of 2.33 A to each heater.

The temperature of the plate was monitored using three Type K thermocouples[†], which were press fit into 0.16 cm-diamter holes drilled from the back side of the plate to within 0.32 cm of

---

[*] Certain commercial equipment, instruments, or materials are identified in this document. Such identification does not imply recommendation or endorsement by the National Institute of Standards and Technology, nor does it imply that the products identified are necessarily the best available for the purpose.

[†] As explained in Section 5, the estimation of standard uncertainties (for time to glowing or flaming ignition and for maximum observed heat release rate) requires a greater number of experiments than could be performed within the scope of this project. In the absence of such data a qualitative approach was adopted in which the degree of collapse of experimental data onto well-defined curves was adopted as a qualitative basis fro assessing uncertainty; therefore, this report does not provide estimates of standard uncertainties, and graphs do not display error bars.

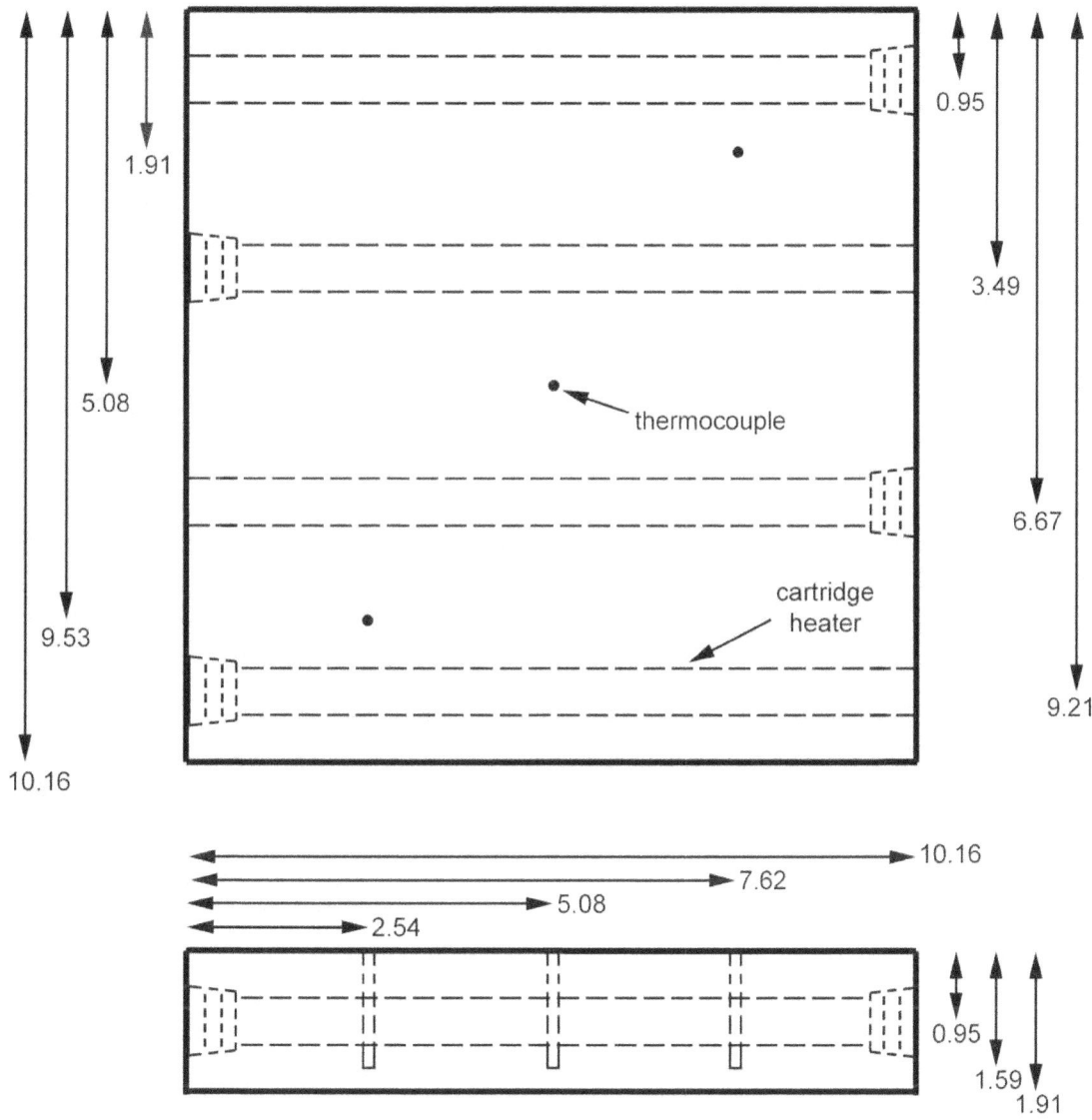

Figure 1. Schematic plan and side views are shown for the heated copper plate used for the ignition studies. All dimensions are in cm.

the top heated surface. Two of the mounting holes were located on opposite corners of the plate at distances of 1.91 cm from the edges parallel to the heaters and 2.54 cm from the edges perpendicular to the heaters. The third thermocouple was located at the center of the plate. The thermocouple leads were connected to Omega DP41-TC thermocouple readers that provided visual readouts of the temperatures at the three locations.

The heated plate was placed on a ring stand. Figure 2 shows a photograph of the plate with the heaters and thermocouples in place. Initial testing of the heated plate in ambient air showed that setting the two power supplies to the same voltage resulted in a current flow for the larger supply

Figure 2. This photograph shows the heated copper plate used for ignition measurements of cellulosic fuels in place on a ring stand. The ends of the four cartridge heaters and their leads can be seen on the two edges of the plate. The leads for three thermocouples imbedded in the plate are visible below.

that was three times larger than recorded for the smaller. This is the expected behavior if the four cartridge heaters have the same resistance values. As the temperature of the plate was increased, the three thermocouple outputs provided temperature readings that agreed within 1 °C, indicating good uniformity of the temperature field within the plate. For recording purposes, the temperature of the thermocouple at the center of the plate was used.

Figure 3 shows the plate temperature recorded as a function of the current applied to each of the cartridge heaters as solid circles from an initial calibration. The result of a quadratic least squares curve fit is shown as a solid line for comparison purposes only. Over the course of the experiments small variations in temperature for a given current were observed that were most likely due to changes in the emissivity of the copper surface. Some changes in the physical appearance of the plate were apparent when the operating temperatures were varied.

3.1.1.2. Air Flows

Experiments were performed with and without air flows over the heated plate. Air flows were generated using a Dayton Model 4C443A blower delivering a nominal volumetric flow rate of

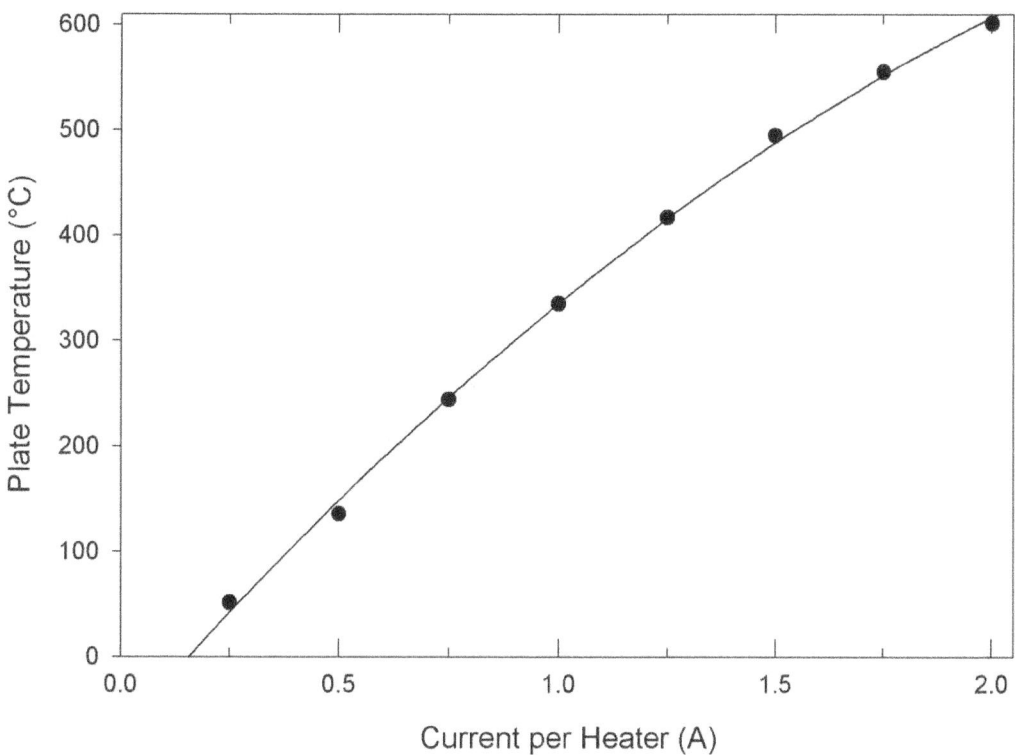

Figure 3. The temperature measured at the center of the heated copper plate is plotted as a function of the current applied to the four cartridge heaters embedded within the plate. Filled circles are the experimental measurements and the solid curve is the result of a quadratic least squares curve fit to the data.

219 L/s. The output of the blower was mounted at the same height as the heated plate and directed towards the plate from a distance of 134 cm. The air velocity at the plate was characterized by recording the output from an Applied Technologies, Inc. Model SPAS/24 sonic anemometer with a portable computer. The anemometer measures wind velocity in two dimensions. The probe was placed 1.2 cm above the plate such that one of the velocity components was aligned along the primary direction of the flow generated by the blower, as shown in Figure 4.

Figure 5 shows an example of the wind speed recorded by the sonic anemometer. Initially the blower was off. When the blower was turned on the air speed increased rapidly to values on the order of 2.5 m/s. Rapid fluctuations in speed are evident during the just over 300 s the fan was on. When the fan was turned off the speed dropped rapidly to that characteristic of the quiescent laboratory.

Averaging the data plotted in Figure 5 yielded means and root mean square (rms) values of 0.09 m/s ± 0.04 m/s and 2.5 m/s ± 0.3 m/s for the wind speeds with the fan turned off and on, respectively. The low wind speed recorded with the fan off is reasonable since the plate was located under an open exhaust hood. The mean wind speed recorded with the fan turned on should be indicative of the applied wind speed, but it likely that the fluctuations are underestimated due to the averaging inherent in recording over the 15 cm path length between

Figure 4. Photograph showing the sonic anemometer probe used to characterize the applied wind velocity in place 1.2 cm above the unheated copper plate.

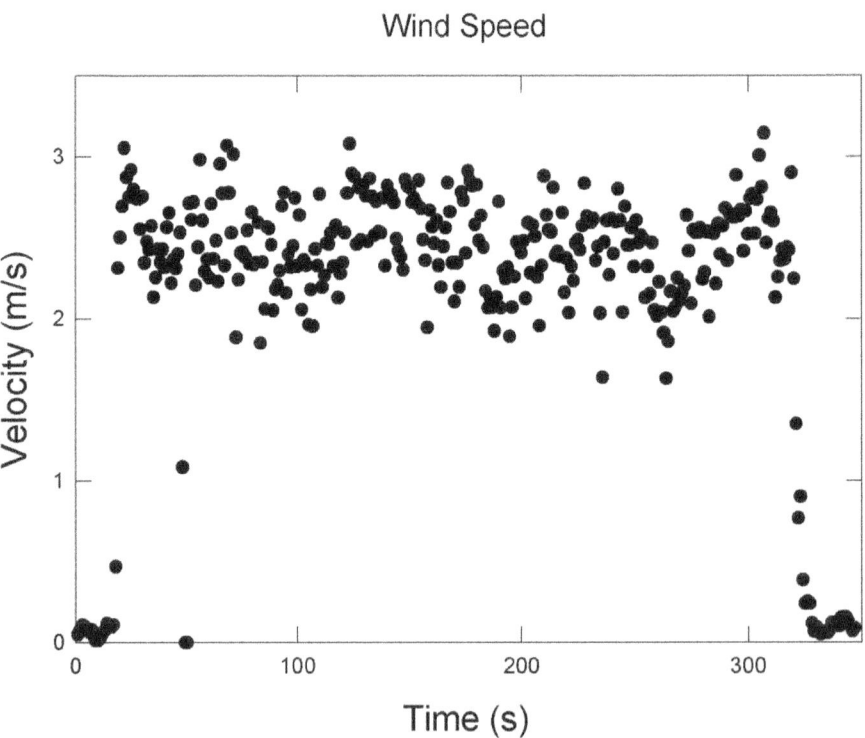

Figure 5. The wind speed above the copper plate along the primary flow direction of the blower recorded as a function of time is shown with the blower turned on and off.

Figure 6. The blower with screens in place used to generate the "low" velocity air flow is shown.

the sonic transmitter and receiver. A more accurate value for the rms would have required measurements with much higher spatial resolution, which were not done.

Velocity measurements such as those shown in Figure 5 were repeated six times over a two month period. The average of the six mean measurements with the fan turned on was 2.53 m/s with individual standard deviations comparable to that for the data in Figure 5. The measured ratios of rms to the average varied from roughly 8 % to 11 %.

A wind speed of 2.5 m/s corresponds to 5.6 mph. It became clear early in the project that lower wind speeds affected the ignition behaviors of fuels placed on the heated plate when compared to cases without wind, and that it was desirable to apply a wind with a lower speed. Instead of generating the lower velocity by controlling the blower output, the velocity was reduced by leaving the flow unchanged and placing a series of four screens with an open aperture of roughly 5.3 cm × 11.4 cm in the flow path. The screens were placed on a holder that could be easily placed into and removed from the blower stream in a repeatable manner by attachment with a clamp to a horizontal rod supported by a vertical mount. Figure 6 shows a photograph of the blower with the screens in place.

The wind speed with the screens in place was measured in the same way described above. Four measurements yielded an average value of 1.1 m/s. The rms values for individual measurements varied from 13 % to 20 %, somewhat higher than observed for the higher speed flow. As discussed above, it is likely that the measurements underestimate the actual fluctuations.

In this report experiments with an applied wind of roughly 2.5 m/s will be referred to as "high wind" cases, while those using 1.1 m/s will be denoted as "low wind" cases. Note that these designations simply refer to the relative magnitudes of the flows used in this study and have no relevance to potential magnitudes of outdoor winds.

3.1.1.3.   Wire-Screen Cage

The purpose of these experiments is to characterize the ignition behavior of typical outdoor cellulosic fuels when placed in contact with a heated surface. Common outdoor fuels such as grasses and pine needles consist of small pieces that form porous fuel beds. These materials had to be contained in some manner in order to create fuel beds that could be reproducibly brought in contact with the heated plate. It was viewed as desirable to allow the fuel beds to have free access to air from the sides and top. As a compromise for these competing requirements, the fuels were contained within a cage fabricated from stainless steel wire screen. The screen was formed from wires with a diameter of 0.71 cm and a wire spacing of 0.32 cm.

The wire-screen cage was designed so that the fuel beds would have nominal dimensions of 10.2 cm × 10.2 cm × 2.5 cm when the cage was placed over the heated plate. An open-bottom wire-screen cage was formed by starting with a 10.2 cm-square section of the wire screen that served as the top of the cage and attaching four wire-screen sidewalls with fine wire. Three of the sidewalls had widths of 10.2 cm and depths of 4.2 cm. Two slots that lined up with the threaded heater mounts visible in Figure 2 were cut in each of the two opposed sidewalls. These slots were wide enough to just allow the cage to slide down pass the mounts while the slot depths of 1.1 cm provided an open height above the plate of 2.7 cm when the cage rested on the mounts. The fourth side wall of the cage was formed from a 10.2 cm wide length of screen with a depth of 2.5 cm. When placed on the cage there was a narrow open gap of approximately 0.013 cm between the bottom of the short side wall and the top of the plate.

In order to apply the fuel within the wire-screen cage to the heated plate it was necessary to provide a bottom of the cage that could be removed. The bottom section that served this purpose is shown in place on the inverted wire-screen cage in Figure 7. This section was cut from a 0.008 cm-thick steel plate. The rectangular portion of the plate has dimensions of 10.2 cm ×10.8 cm and was designed to slide between the two deep sidewalls on opposite sides of the cage. The 2.54 cm × 1.27 cm tab on the right side of the bottom section in Figure 7 slides through a slot cut 2.5 cm from the top of the third deep sidewall and rests on top of the short sidewall on the opposite side of the cage. The narrow section on the left side served as a handle that allowed the bottom plate to be slid along the top of the plate and removed to expose the fuel to the plate.

The fuel to be tested was placed in the upside down wire screen cage to a depth of 2.5 cm. The bottom plate was then inserted into the slot as shown in Figure 7. By grasping the handle, the wire-screen cage was carefully inverted and place over the heated plate. The bottom was then rapidly removed by sliding it away from the cage. Note that the bottom plate insulated the fuel from the heated plate until it was rapidly removed, thus providing a well defined time of fuel application.

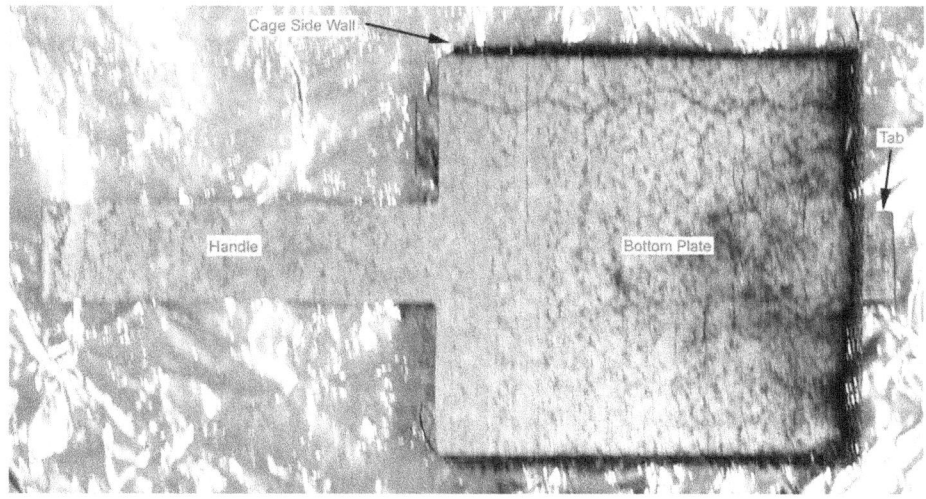

Figure 7. The wire cage with its removable steel-plate bottom in place is shown looking down from above.

It should be noted that the wire-screen cage provided resistance to the air flow into the fuel bed that varied with the velocity passing through the screen. A test with a single flat section of the screen placed upstream of the test plate indicated that the high wind flow velocity was reduced by 32 % by passing through the screen.

3.1.2. Visual Recordings of Ignition Experiments

The principle experimental diagnostic used for the heated plate ignition experiments was videography. Each ignition experiment was recorded using a Sony Model DCR-PC100 mini-DV video camera. The taped results were subsequently analyzed to determine the time when the fuel was placed on the plate and the times when glowing combustion and flaming were observed. Other behaviors such as the location and amount of smoke released could also be tracked. The temperature of the plate was monitored by reading the temperature recorded by the center thermocouple in the plate out loud so that it was recorded on the audio track of the video. Subsequent analysis of the video tape allowed the temperature to be determined as a function of time. Other visual observations were also recorded in this way.

3.1.3. Experimental Procedure

The various fuels investigated during the study are described below. For each fuel the mass of material required to fill the approximately 262 cm^3 volume of the wire-screen cage was determined, and this mass was used for all experiments with that fuel. In filling the cage no attempt was made to order or compress the fuel. The resulting fuel beds were typically loosely packed and porous. The density of the fuel bed was calculated by dividing the fuel mass by the nominal fuel bed volume.

Figure 8. This photograph shows the wire-screen sample cage with its steel-plate bottom in place on top of the heated plate. The cage is filled with cheat grass. The fuel is placed on the plate by withdrawing the bottom of the cage using the handle visible in the photograph.

The pre-determined mass of fuel was then loosely evenly distributed into the inverted wire-screen cage through the open bottom. Once the fuel was distributed, the tab on the plate that formed the bottom of the cage was inserted into the slot in the wire-screen side wall and allowed to rest on the opposite side wall.

The first step in an experiment was to heat the plate to the desired temperature by adjusting the equal currents supplied to each heater by the two power supplies. When an experiment was to be run the camera was first started and identifying information such as the date and test conditions were recorded audibly. The wire-screen cage was then carefully inverted and placed over the heated plate as shown in Figure 8. Within a few seconds the plate was extracted from the bottom of the cage, and the fuel was applied to the plate. If a wind was to be applied, the fan was turned on immediately prior to or just after the application of the fuel.

The application of the fuel and wind changed the heat losses from the plate and adjustments to the currents were required to maintain a constant temperature. For instance, applying the fuel in the absence of the wind typically caused the temperature to increase, indicating that heat transfer to the fuel was more than offset by reduced heat losses from the plate. In this case the currents to the heaters needed to be reduced in order to maintain a constant temperature. On the other hand, applying a wind without adding the fuel caused the temperature to drop due to increased convective heat transfer to the air from the plate surfaces, and it was necessary to increase the currents to maintain a constant temperature.

By trial and error the amount of current change required to maintain a nearly constant temperature for a given fuel and wind condition was determined. As soon as the fuel was applied and the wind (if applicable) was turned on, the current would be manually adjusted to the appropriate level. During the remainder of the experiment the plate temperature would be monitored and small current adjustments would be made as necessary to hold the temperature variation within a small range. The actual current changes were not recorded, but it was noted that there was a correlation between the required current and the behavior of the fuels. For instance, when glowing combustion was observed, it was typical to have to reduce the current slightly in order to maintain the temperature.

Experiments showed that it was normally necessary to reduce the currents substantially when fuels were applied without a wind, relatively small changes in current were required when experiments were done with a low wind, and increased currents were needed when a high wind was used. The changes in temperature were relatively slow, and it was generally possible to hold the recorded temperatures within ± 2 °C. Occasionally larger excursions were observed, but variations were always much smaller than the smallest step size (typically 10 °C) used for plate temperature settings. Figure 9 shows a plot of representative temperature histories recorded for a range of temperature settings and wind condition (none, low, high). Times are generally shorter at higher temperatures since an experiment was halted when flaming was observed or visual observation (e.g., smoke no longer visible) suggested no additional reaction was taking place.

At the conclusion of an experiment the video camera was shut off. At a later time the video tape was replayed in order to record the plate temperature readings as a function of time, ignition times for glowing and flaming combustion, and other experimental observations such as smoke behavior.

3.2. Ignition by Thermal Radiation

Ignition of cellulosic fuels as the result of exposure to thermal radiation was investigated by placing loosely packed samples of the material to be tested in a cone calorimeter operated in a non piloted ignition mode. The exposed surface of the fuel bed was oriented facing upward. The response of the sample to an imposed heat flux was observed while maintaining a constant heat flux. The time required to ignite the fuel was recorded. The dependence of the ignition time on heat flux level for a given fuel was determined by repeating the experiment for a range of imposed heat fluxes. Other parameters available from the experiments, in addition to time to ignition, included time behaviors of heat release rate, mass, and smoke optical density.

3.2.1. Cone Calorimeter

The cone calorimeter is a widely available standard instrument used primarily to measure the heat release rate of small samples when subjected to a uniform radiative heat flux. Heat release rate is measured using the oxygen depletion approach. The cone calorimeter has been adopted

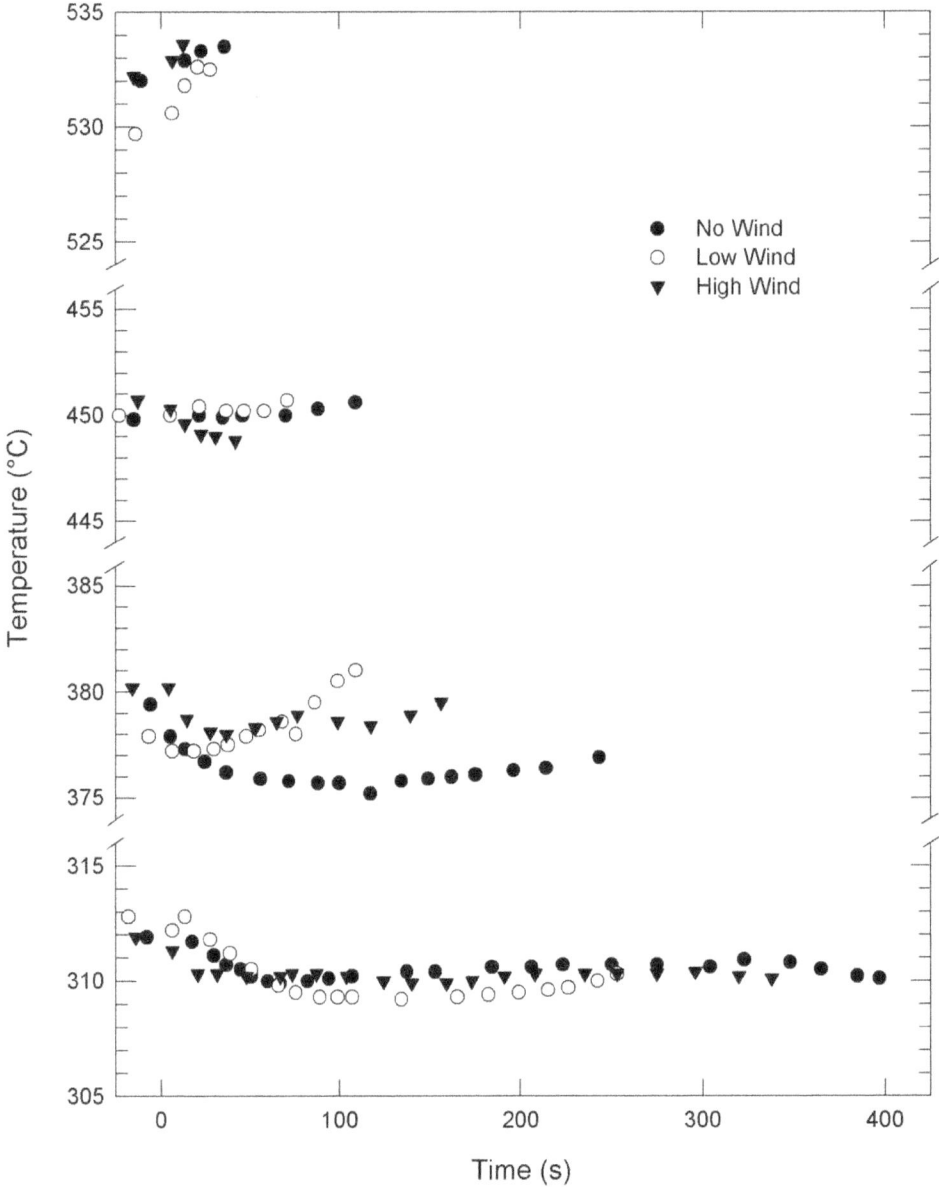

Figure 9.   Temperatures recorded during experiments in which cellulosic fuels were place on a heated plate are plotted as a function of time. Zero time is defined as the time when the fuel was exposed to the surface.

as a standard method by ASTM International (ASTM E-1354) [12], the International Organization for Standardization (ISO 5660) [13], and the National Fire Protection Association (NFPA 264A) [14].

While primarily designed for heat release rate measurements, the cone calorimeter can be used to study radiative ignition since it provides a nearly uniform and easily varied source of radiative

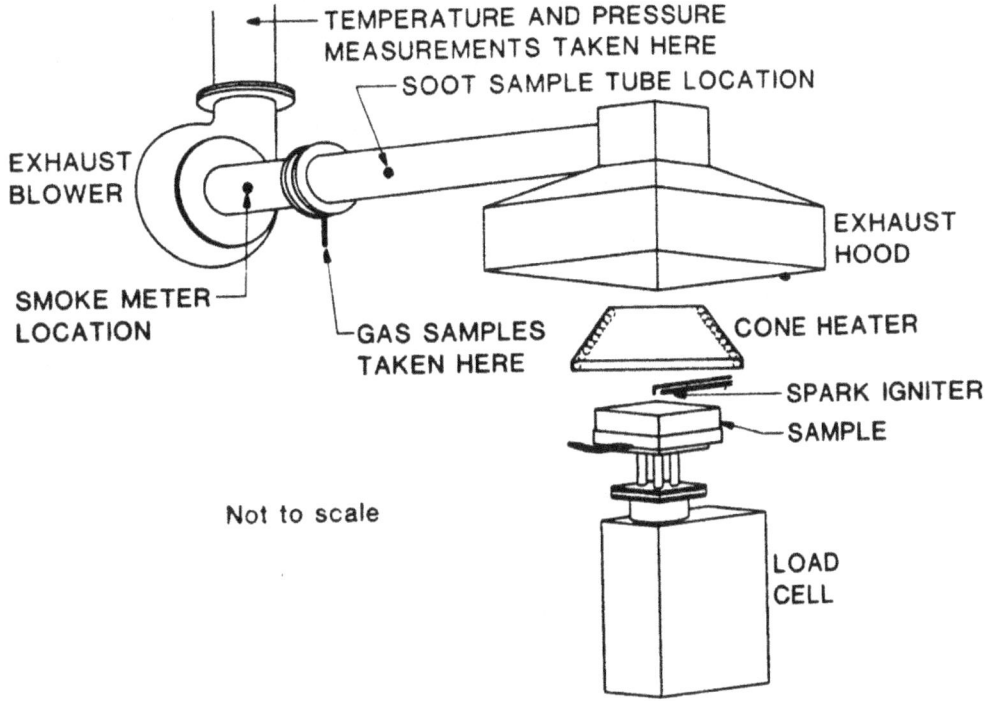

TEMPERATURE AND PRESSURE
MEASUREMENTS TAKEN HERE

SOOT SAMPLE TUBE LOCATION

EXHAUST
BLOWER

EXHAUST
HOOD

SMOKE METER
LOCATION

GAS SAMPLES
TAKEN HERE

CONE HEATER

SPARK IGNITER

SAMPLE

Not to scale

LOAD
CELL

Figure 10. This schematic shows the principal components of the cone calorimeter used to investigate the ignition behavior of cellulosic fuels and polyurethane foam.

heat flux at the sample surface. Babrauskas and Parker describe the use of the cone calorimeter for ignition studies. [9]

Figure 10 shows a schematic diagram of a cone calorimeter. The principal components are a conical resistance heater wound in the form of a truncated cone with temperature control achieved using 3 type K thermocouples as inputs for a controller that automatically adjusts the current supply, cooled shutter to protect the sample before the start of a test, a specimen holder placed on top of a load cell for mass measurement, a spark igniter used for piloted ignition of fuel vapors, a hood and blower driven exhaust system equipped with an orifice plate for velocity measurements, a gas sampling port used to extract gases for oxygen (paramagnetic analyzer) and other molecular species concentration measurement, and a laser system for monitoring smoke obscuration.

The cone calorimeter used for these experiments was manufactured by Atlas Combustion Analysis Systems. It was subsequently modified by replacing the original software for controlling the experiment and performing analysis with software developed by Fire Testing Technology Ltd. The results of measurements include observed ignition times and heat release rate per unit area, sample and sample holder mass, and smoke optical density as a function of time, along with such derived properties as maximum heat release rate, time to maximum heat release rate, total mass loss, effective heat of combustion, etc. The experimental results were imported into spreadsheet files and displayed using a standard plotting software package.

Figure 11. A sample holder used during the radiative ignition experiments is shown filled with fuel.

3.2.2. Experimental Procedure

The experiments were performed while generally following the standard cone calorimeter protocols. Sample sizes were 10.2 cm ×10.2 cm × 2.54 cm and were contained within a sample holder formed from a double-layer of aluminum foil, which was open at the top. Note that unlike the experiments on heated surface ignition, the sides of the sample holder were closed. The sample holders for the plate heating and cone calorimeter experiments were the same size, and the mass of a particular fuel used was also the same. Figure 11 shows one of the sample holders filled with fuel. No external wind was imposed during these measurements, and the only air flow was that induced by the cone calorimeter blower and natural convection from the samples. The separation between the top of the fuel bed and base of the cone was 2.5 cm.

Prior to an experiment the radiative heat flux at the location of the sample was set by adjusting the current to the cone heater until the desired level was measured with a calibrated gauge. At this point the water-cooled shutter was positioned between the cone and the sample area to limit preheating of the sample. After the sample holder and fuel were placed in position and background measurements were completed, the shutter was rapidly removed to expose the sample to the imposed heat flux. This defines time zero. Cone experiments often incorporate a spark igniter above the sample to provide a pilot for the flammable gases that are generated from the fuel. Since the purpose of these experiments was to investigate non piloted ignition, the igniter was withdrawn from above the sample and was not activated during these measurements.

Ignition times were determined visually by noting the appearance of either glowing combustion or flaming. In practice, the measured heat release behaviors also provided good indications for ignition times. An experiment was allowed to continue until obvious changes in fuel appearance were no longer apparent and the time rates of change for mass and heat release rate were small.

3.3. Fuels Investigated

A variety of cellulosic fuels including shredded newsprint, four grasses, and pine needles were investigated. In addition, limited measurements were carried out on three types of dried leaves. The ignition behavior of a common plastic fuel, non fire-retarded flexible polyurethane foam, was investigated in the cone calorimeter.

Free moisture absorbed in the fuel is known to be an important parameter for the ignition and burning behavior of cellulosic fuels. No attempts were made to control free moisture, defined as the percentage mass of water relative to the dry material, in this study beyond allowing the fuels to come into equilibrium with the surrounding laboratories, which had nominal temperatures of 20 °C with variable relative humidity that did not exceed 50 %. Free moisture in the fuels was measured using the procedure described by Manzello et al. [22,23] in which the material to be tested is placed in an oven held at 105 °C for three hours and the amount of mass loss quantified. These authors showed that no further changes in mass occurred after 3 hours for these thin fuels.

Several grams of the material were placed in a small tared aluminum pan and weighed to ± 0.0001 g. The pan and grass were then placed in the oven. After three hours the pan was removed from the oven and allowed to cool in a desiccator. The cooled fuel sample and pan was quickly reweighed after removal from the desiccator in order to limit the uptake of moisture by the dried sample. The free moisture percentage was calculated using the following equation,

$$\frac{m_{f,m} - m_{f,d}}{m_{f,d} - m_p} \times 100, \tag{1}$$

where $m_{f,m}$ is the measured mass of the pan and fuel with moisture, $m_{f,d}$ is the mass of the pan and dried fuel, and m_p is the mass of the empty pan.

3.3.1.   Shredded Newsprint

Commercial newsprint was obtained from a local publisher as remnants of large 61.0 cm wide rolls. Measurements with a micrometer at several locations on a small area of the newsprint yielded a thickness of 6.4×10^{-3} cm. Sections of the newsprint were run through a shredder that generated strips that were roughly 0.4 cm wide by 4 cm long. There were substantial variations in the lengths of individual pieces.

Fuel beds were created by loosely packing 6.9 g ± 0.1 g of the shredded newsprint into the wire-screen cage sample holder. Figure 12 shows a photograph of the shredded newsprint in the wire-screen cage.

The free moisture percentage of the shredded newsprint was measured well after ignition testing was completed. Two independent measurements yielded values of 6.2 % and 6.4 %.

Figure 12. Photograph of shredded newsprint in wire-screen cage.

Figure 13. Photograph of May tall fescue in wire-screen cage.

3.3.2. May Tall Fescue Grass

A sample of tall fescue grass (*Festuca arundinacea*) was collected on the NIST campus (GPS location of 39.08.082 N, 77.12.709 W) on May 25, 2006. The site was the outfield of a softball field that was planted in tall fescue about six years ago. The grass had been cut within the past two days. Clippings that had been cut and expelled from the mower and were resting on top of the growing grass were collected. The grass clippings felt dry and light to the touch. The weather over the previous few days had been nearly ideal for drying with no rain, a temperature range of roughly 7 °C to 21 °C, and dew point temperatures on the order of 4 °C.

Spring is a time of rapid grass growth in this region. Significant rain had fallen within the past two weeks. The uncut grass was thick and dark green. As a result of the conditions, the grass had been fairly long when cut, with the clippings having a range of lengths varying from 8 cm to 15 cm. A small amount of the grass was topped with seeds. Even after drying the grass clippings retained a light greenish color. Much of the collected grass consisted of thin blades.

After several weeks of storage in a general purpose laboratory, three independent measurements of free moisture percentage taken over several days yielded values of 12.8 %, 11.9 %, and 13.6 %. These values are typical of those expected when the moisture in the grass has come into equilibrium with air in the surrounding laboratory.

As collected, the tall fescue did not pack well in the sample holders due to its length. Scissors were used to cut bunches of the grass into lengths of 2.5 cm to 5 cm, which were used to form the fuel beds. A mass of 9.0 g of material was found to be adequate for forming a 2.5 cm deep bed when lightly packed into the sample holders. Figure 13 shows a sample of tall fescue grass in the wire-screen cage used for the heated plate experiments. For the remainder of the report these samples will be referred to as "May tall fescue."

Figure 14. Photograph of August tall fescue in wire-screen cage.

Figure 15. Photograph of cheat grass in cone calorimeter aluminum pan.

3.3.3. August Tall Fescue Grass

A second sample of recently cut tall fescue grass was collected from the same location as the May tall fescue on August 29, 2006. Samples of this grass will be referred to as August tall fescue. Once again, clippings created by mowing were collected from above the uncut grass.

August in this region is typically hot and often dry. Weather records indicate that over the four prior weeks daytime high temperatures ranged from about 27 °C to 38 °C, with lows ranging from 13 °C to 27 °C. The last significant rain had fallen on August 7th.

This grass had a very different appearance than that collected in May. It was tan colored and appeared to consist primarily of stalks as opposed to blades. Its length varied from about 2.5 cm to 7.5 cm, and it could be packed into fuel beds without cutting.

Figure 14 shows a fuel bed formed from 7.0 g of August tall fescue.

The grass felt dry to the touch when collected. A measurement of its free moisture percentage after two days in the laboratory yielded a value of 13.1 %.

3.3.4. Cheat Grass

Figure 15 shows one of the samples of cheat grass in an aluminum pan. Cheat grass (*Bromus tectorum*) is an invasive plant introduced into the United States in the 1890s. It has subsequently spread widely across the western United States. Cheat grass is a winter grass, growing rapidly in late winter and spring. In summer it dies off, leaving a dried grass that is widely regarded as highly flammable and a major contributor to wildland fires.

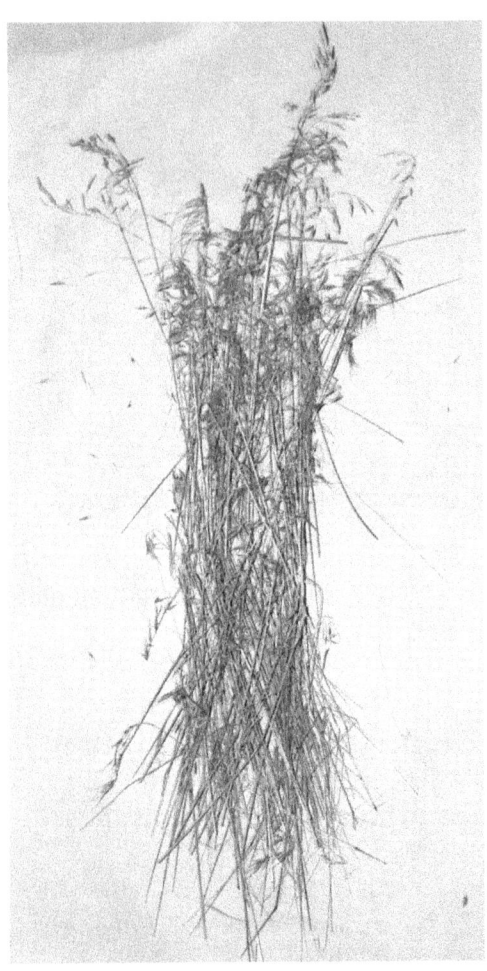

The cheat grass for this investigation was provided by researchers from the United Stated Department of Agriculture, Agricultural Research Service Laboratory in Reno, NV. The grass was collected in the Honey Lake Valley of northeastern California (GPS coordinates 40.08.291 N, 120.04.672 W) on July 19, 2006 and shipped to Gaithersburg, MD. During the week prior to collection high temperatures were around 38 °C with relative humidity below 10 %.

Figure 16 shows a portion of the cheat grass as received. The grass was cut near its base and had heights on the order of 45 cm. The grass felt dry to the touch and was golden in color. It consisted of very slender stalks topped with branching seed structures known as panicles. The free moisture percentage of a sample of the grass was measured to be 11.2 %.

As received, the grass could not easily be packed into the fuel beds. The grass was cut into lengths of 2.5 cm to 5 cm using scissors. No special care was taken to isolate the stalks and panicles, and both were present in the fuel beds. A mass of 8.0 g was sufficient to fill the sample holders.

Figure 16. A photograph showing an uncut sample of cheat grass

3.3.5. Fine Florida Grass

A grass sample collected at a lawn mower manufacturer's test site near Tampa, FL was provided for the study. The grass had been removed from the tops of mower decks and was intended for testing the debris that collects on mowers. The sample had been collected at least a year earlier and was stored in bags under dry conditions. The grassy material was cut finely and felt fluffy when unpacked, having a consistency similar to sawdust. The material was brown in color. No additional information was available, including the type of grass that had been cut or weather conditions. This material will be referred to as fine Florida grass.

An 8.0 g sample of the fine Florida grass just filled the wire-screen cage. Figure 17 shows a photograph of the cage filled with fire Florida grass. Two measurements of fuel moisture percentage recorded well after the ignition experiments were completed yielded values of 10.1 % and 9.6 %.

Figure 17. Photograph of fine Florida grass in wire-screen cage.

Figure 18. Photograph of pine needles in wire-screen cage.

3.3.6. Pine Needles

Pine needles were obtained from a commercial operation (Florida Pine Straw Company of Mayo, FL) that distributes pine needles for use as mulch. These needles come from loblolly pine trees (*Pinus taeda*). They are collected by machines that sweep up pine needles that have fallen to the ground. They were delivered in bales that were opened and left exposed to the laboratory.

The pine needles were dark brown and dry to the touch. The fuel moisture was measured to be 13.7 %. They had lengths that generally varied from 15 cm to 23 cm with rectangular cross sections having rough dimensions of 0.5 mm × 1.4 mm (dimensions were highly variable). Individual needles were brittle and easily broken. Often two or three pine needles were connected together at their base by a needle sheath, forming a fascicle. Since the needles were typically too long for the sample holders, they were cut into lengths between 2.5 cm and 5 cm using scissors for testing purposes. No attempt was made to separate the sheaths from the needles. Small debris, most often small pieces of pine wood, was removed when encountered.

The cut pine needles formed a dense bed, with 16.0 g used to fill the sample holders. Figure 18 shows a photograph of a bed of pine needles in the wire-screen cage.

3.3.7. May Tall Fescue/Pine Needle Mixture

A mixture of May tall fescue grass and pine needles was tested using materials processed as described above. The masses required to fill the cage for the grass (9 g) and pine needles (16 g) were quite different. The masses of each required for the mixture were calculated for a 50 %/50 % mixture by mass assuming the volumes of each component would be independent of mixing. The result was 5.8 g for a total mass of 11.6 g, with grass making up 64 % of the volume and the pine needles 36 %. The two materials were weighed independently and then mixed by hand until

Figure 19. Photograph of the May tall fescue/pine needle mixture in the wire-screen cage.

Figure 20. Photograph of polyurethane in a cone calorimeter sample holder.

they appeared to be evenly distributed. The mixture was then placed in the wire-screen cage. Figure 19 shows a photograph of the resulting mixture.

3.3.8. Preliminary Tests with Leaves

A limited number of heated-plate experiments were done using leaves collected locally. These included leaves from boxwood (*Buxus*), American elm (*Ulmaceae Ulmus Americana*) and pin oak (*Quercus palustris*).

3.3.9. Polyurethane Foam

Flexible polyurethane foam was tested in the cone calorimeter. The samples were obtained from a commercial supplier and were not fire retarded. The exact composition of the material is not known. The sample thickness was 5.1 cm. Sample masses varied over a range of 11 g to 12 g. Figure 20 shows a photograph of a sample of the polyurethane surrounded by the aluminum foil used as the holder for the cone experiments.

3.3.10. Fuels Summary

Table 1 includes the mass used, nominal density, and measured moisture content for all of the fuels used during the current investigation. Missing data represent cases for which the amount of fuel available was limited or measurements were not made.

Table 1. Summary of Fuels, Sample Mass Used, Nominal Fuel Bed Densities, and Measured Fuel Moisture

Fuel	Sample Mass (g)	Nominal Density (g/cm^2)	Fuel Moisture (%)
Shredded Newsprint	6.9	0.026	6.2 6.4
May Tall Fescue	9.0	0.034	12.8 11.9 13.6
August Tall Fescue	7.0	0.027	13.1
Cheat Grass	8.0	0.031	11.2
Fine Florida Grass	8.0	0.031	10.1 9.6
Pine Needles	16.0	0.061	13.7
May Tall Fescue/Pine Needle Mixture	11.6	0.044	--
Boxwood Leaves	8.0	--	--
American Elm Leaves	5.5, 5.5	--	--
Pin Oak Leaves	6.0, 9.4	--	--
Polyurethane Foam (5.1 cm thickness)	11 to 12	0.042 to 0.046	--

27

Figure 21. The smoke plume rising from a pine needle fuel bed in contact with the heated plate held at a temperature of 380 °C is shown 213 s after application.

4. Results

4.1. Some General Observations from Heated Plate Ignition Studies

A variety of qualitative observations provide insight into the characteristics and response of the porous fuel beds when placed on a heated plate. First, cases without an applied wind will be discussed. This will be followed by observations when low and high winds were used.

When the temperature of the heated plate was high enough, all of the fuels tested generated a white "smoke" that would flow from the top of the fuel bed. The composition of the smoke is unknown, but it is likely that it contained varying proportions of condensed water vapor and organic pyrolyzate. At times, particularly after it first appeared, the smoke would rise from localized sections of the fuel bed surface. More commonly, the smoke would appear over a large fraction of the surface area. In the latter case, the smoke immediately above the fuel bed typically assumed a circular shape such that there was a smoke-free area around the perimeter of the square fuel surface. Above the surface, the smoke would rise and its area would decrease so that the smoke near the surface adopted a conical shape. As the smoke moved further away from the plate it would develop the fluctuating shape characteristic of a buoyancy-dominated fluctuating flow. Figure 21 shows an example of this type of smoke flow for a bed of pine needles placed on the plate held at 380 °C.

Figure 22   Two frames taken one second apart from a video of a hot surface ignition experiment with May tall fescue are shown. A small area of glowing combustion had appeared during the one second period (compare the blown up areas prior to (left) and after (right) ignition).

The appearance of smoke at the top surface of a fuel bed indicates that the bed is porous to the buoyant flow of heated gases generated within the bed by contact with the heated plate and any heat release taking place within the fuel bed. The characteristic shape of the smoke at the fuel surface suggests that the buoyant flow within the bed acts much like a buoyant plume in open air, entraining air radially inward that has passed through the sides of the fuel bed.

The amount of smoke released from a fuel bed (qualitative description, estimated visually) varied widely with time, the fuel tested, and plate temperature. The level of smoke observed ranged from just barely visible to quite heavy as seen in Figure 21. It was clear the smoke density was related to the amount of fuel pyrolysis that was taking place. Particularly for the lower plate temperatures where relatively long periods were required for glowing combustion and/or flaming to appear, the smoke would tend to increase with time, becoming quite heavy before glowing or flaming appeared.

During the experiments, the times for the appearance of glowing combustion and flame were noted visually and recorded audibly on the video tape. The times were determined a second time based on a visual review of the video tape. The video camera used for the investigation responds to light in the near infrared, and was particularly useful for detecting the onset of glowing combustion as long as it is not hidden from view within the fuel bed. For experiments without an applied wind, glowing combustion was usually observed on the video tape before it was seen visually. The sensitivity possible using the video camera is demonstrated in Figure 22, which compares two frames taken one second apart from a video of an ignition experiment for May tall fescue with a plate temperature of 381 °C and no wind applied. The appearance of the glowing combustion is easily identified by comparing the two enlarged areas. Glowing combustion would generally appear at a small single point in the fuel bed and then spread to cover larger areas.

The glowing combustion was not generally hidden within the fuel bed for cases without an applied wind because there was a high likelihood that it would appear on the side of the fuel bed facing the camera. The most likely reason for this side to be favored was the narrow open slit

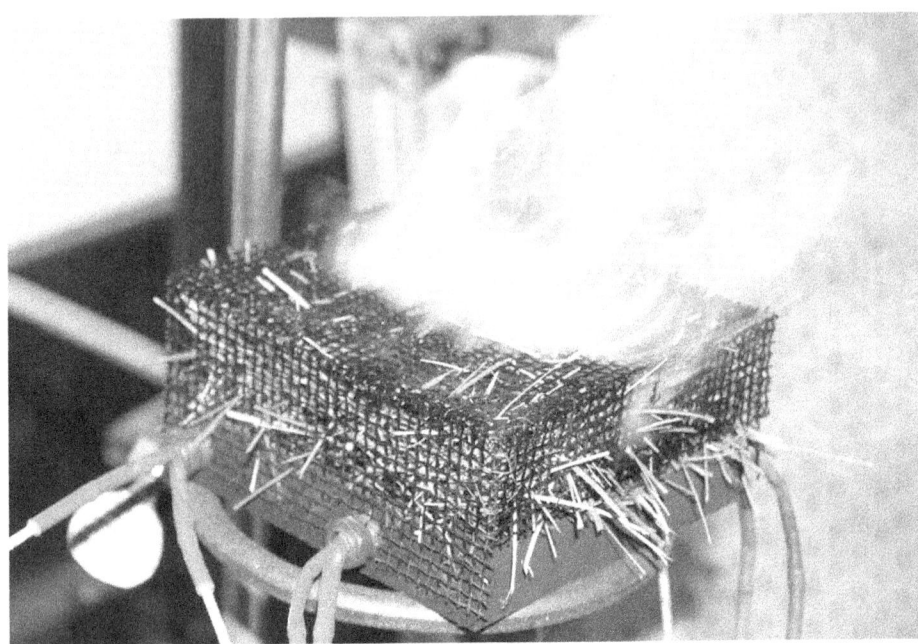

Figure 23.   The smoke generated by a mixture of May tall fescue and pine needles applied to the heated plate at 380 °C in the presence of a low wind is shown 175 s after application.

between the heated plate and the base of the wire-screen sidewall. As discussed earlier, smoldering development is very dependent on the amount of oxygen reaching the fuel surface. Apparently, the small slit allowed more air to enter the cage from this side than passed through the screens on the other three sides, with the result that the initial surface heating required to ignite glowing combustion was more likely near the slit. The practical effect of this preference is likely a short decrease in the time required for glowing combustion to develop compared to that which would have been observed if the screen wall extended below the heated plate.

The observations changed substantially when a wind was applied. For both low and high wind cases, smoke that came from the top of the fuel bed tended to lie along a narrow band oriented perpendicular to the flow direction. Figure 23 shows an example of this behavior for the May tall fescue/pine needle mixture with a plate temperature of 380 °C and a low wind applied. This photograph was taken 175 s after the fuel was placed on the plate. Note that the band of smoke across the top of the fuel bed also extends down the side of the cage.

While the amount of smoke generated varied with fuel type, surface temperature, and time after fuel application, the formation of a smoke band was normally observed with an applied wind. The location of the smoke band relative to the upstream edge would often shift somewhat during an experiment. It was also observed that for a given fuel type, the band would be located further downwind for the higher wind case. The upstream edge of the band tended to appear anywhere from upstream of the center of the fuel bed to locations reaching at least 4/5 of the way across from the upstream side.

As the wind passed through the upstream side of the wire-screen cage, it was certainly attenuated in the manner shown experimentally above. The porous fuel bed would slow down the air flow further. It is the interaction of this attenuated wind with the buoyant flow rising from the heated plate and fuel that leads to the formation of the smoke band. The higher velocity flow is able to penetrate further, thus explaining the downstream shift observed for the higher wind velocities.

The presence of an air flow also affected observed ignition behaviors. The wind causes two competing effects that can affect surface ignition. On one hand, the flow convectively cools the fuel and reduces fuel surface reaction rates. On the other hand, it substantially increases the amount of oxygen reaching the fuel surface, which enhances surface reaction rates. For most of the experiments described below, the latter effect was dominant since both glowing and flaming combustion appeared at locations towards the upstream sides of the fuel beds.

One result of applying a wind was to slightly increase uncertainties associated with measurements of glowing ignition time. Unlike for no-wind cases, glowing ignition was often observed at locations well away from the fuel bed side facing the camera. This is another indication that wind is enhancing surface reactions, since it suggests it is overwhelming the effect of increased air flow through the narrow gap in the wires-screen cage on the camera side. The nascent ignitions with a wind were often hidden from the camera by the fuel bed, and it was common for the earliest observation of glowing combustion to be made visually by the observer. This effect is likely exacerbated when the ignition is location further within the fuel bed away from the observer. It is difficult to estimate how much earlier glowing combustion may have been present before it was detected, but it is likely to have involved relatively short periods since glowing combustion tended to spread rapidly in the presence of a wind, making it more apparent to both the observer and the video camera.

Once glowing combustion appeared, it tended to spread more rapidly and become more intense than in the absence of wind. Generally, the glowing would move downstream away from the location where it was first observed. Often a band of unburned fuel was observed along the upstream edge of the bed. Strong temporal oscillations in glowing intensity with periods on the order of a few seconds were frequently observed. It was unclear if these oscillations were due to variations in the wind velocity or some instability associated with the combustion behavior.

Some qualitative observations concerning the development and behavior of glowing combustion and the transition to flaming both with and without wind are relevant. Glowing combustion tended to appear at isolated small areas, here referred to as pinpoints, similar to those visible in Figure 22. In the absence of a wind the glowing pinpoints could usually be observed moving slowly over individual pieces of fuel. This seems to imply that glowing combustion is a distinct process from non glowing combustion that is self sustaining. As noted above, glowing is generally believed to be due to surface oxidation of char that had been formed by earlier fuel pyrolysis.

In the presence of a wind, the intensity of glowing was enhanced substantially and at times would appear to be more wide spread than the pinpoints typically observed without wind, with glowing over longer lengths of individual fuel elements.

A glowing fuel element appeared to be capable of inducing glowing in nearby fuel elements. As a result glowing combustion usually spread away from locations where it was initially noted. In this way volumes of glowing combustion would develop. Since the number of glowing fuel elements was increasing, the amount of heat release associated with glowing would increase during these glowing periods.

In numerous experiments the transition to flaming was captured on video. A blue flame would appear over a small volume within the fuel bed and then spread rapidly to other locations. At times, these flames moved with speeds on the order of tens of cm/s. The location where the initial flame appeared was always surrounded by an area of extensive glowing. Within a short period of time yellow flames would appear, and the flaming became self sustaining.

The observations concerning transition to flaming suggest that gaseous combustion developed in regions within the fuel beds where a flammable premixed gaseous fuel/air mixture had accumulated and eventually ignited. A rapidly spreading blue flame is characteristic of a premixed flame. The gaseous fuel was generated by pyrolysis of the solid fuel and mixed with air diffusing or flowing into the fuel bed. The initial rapid spread of the flames indicates that flammable mixtures were present over extensive volumes. Heat and/or free radicals are necessary to ignite a premixed fuel/air mixture. The fact that ignitions were observed in areas of intense glowing suggests that sufficient heat was present to ignite the locally flammable gas mixture. While not conclusive, the observation that transition to flaming occurred in areas of intense glowing suggests that glowing is required to ignite flaming in these pyrolyzing fuel beds.

Once a premixed flame developed, the increased heat release rate rapidly pyrolyzed additional fuel, which then burned as a diffusion flame. The appearance of yellow flames, due to soot formation and subsequent irradiance from within the flame, is indicative of diffusional flame burning.

4.2. Shredded Newsprint

Both heated plate and cone calorimeter ignition experiments were done for this fuel.

4.2.1. Heated Surface Ignition Results

Figure 24 shows heated-plate results for the times required for the onset of glowing combustion and flaming as functions of surface temperature and wind condition for shredded newsprint. The symbols and protocols used in this figure will be used throughout the remainder of the report. The wind condition is indicated by both symbol and color as follows: no wind—red circle, low wind—green square, and high wind—blue triangle. Filled symbols represent results for glowing combustion, while open symbols are used for the onset of flaming. As the temperature was lowered, temperatures were eventually reached where neither glowing combustion nor flaming were observed after long periods. These experiments were allowed to proceed until it became clear (no indication of additional reaction) that ignition was unlikely. Instead of showing the actual experimental times, these results are indicated by plotting the appropriate filled symbol near the upper temperature axis.

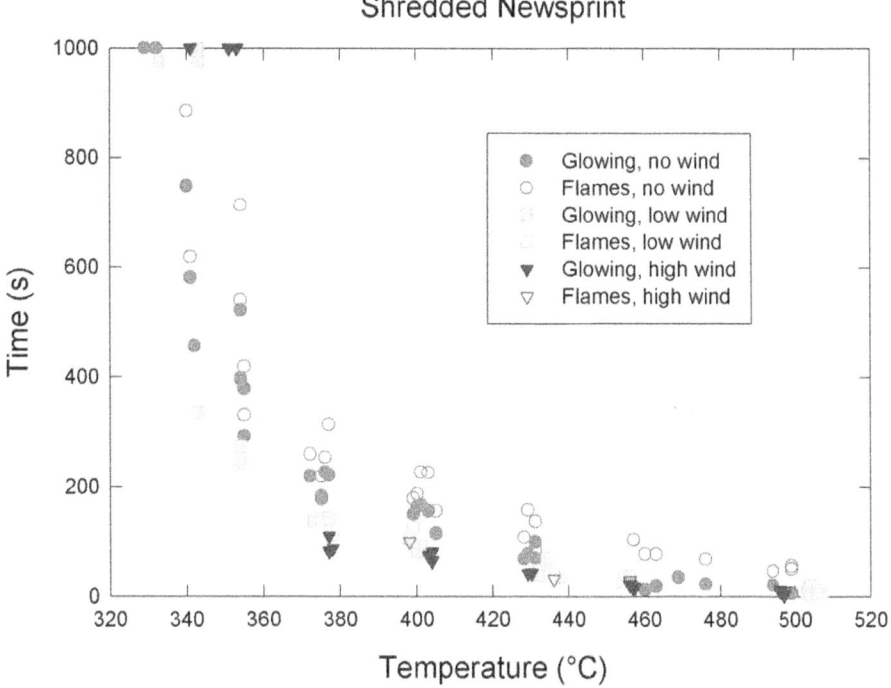

Figure 24. Ignition times for glowing combustion and flaming are shown as a function of temperature for shredded newsprint fuel beds applied to a heated plate. Results are included for no wind, low wind, and high wind cases.

While there is some scatter in results, clear trends emerge from the measurements shown in Figure 24. As expected, the times required for the appearance of glowing combustion and flaming increased as the plate temperature decreased, with data for all of the conditions falling on well defined curves. For temperatures around 500 °C glowing ignition appeared within a few seconds. With a wind applied the glowing combustion rapidly transitioned to flaming, while periods ranging from 26 s to 50 s were required when no wind was present. The time required for the appearance of glowing combustion increased to more than 400 s at temperatures around 340 °C. One of the three tests around this temperature did not transition to flaming, but the other two did, requiring 38 s and 137 s after glowing first appeared. A review of the data shows that for the 30 experiments where glowing was observed without an applied wind, only four failed to transition to flaming. These four had heated plate temperatures of 342 °C , 375 °C, 429 °C, and 469 °C. The wide temperature range suggests that the failure to transition was somehow dependent on a property of the fuel and not directly related to the surface temperature.

For temperatures between 380 °C and 500 °C the effect of applying a wind was to reduce the times required for glowing ignition to appear as compared to cases without wind. These reductions approached 50 % for the lower part of this temperature range. The reductions were greater with a high wind than with a low, though the reductions were smaller than those between the no-wind and low-wind cases.

The time required for transition to flaming was much shorter with wind applied than for the no-wind cases. In all cases with an applied wind, if the shredded newsprint began to glow, it transitioned to flaming in a short period of time.

When the heated plate temperature was lowered to around 350 °C glowing combustion was no longer observed in the presence of the high wind, but ignition still occurred for the low wind and no wind conditions. Even though glowing combustion did not develop, there were indications that some pyrolysis of the fuel did occur. Smoke was observed rising from the fuel bed within a few seconds of being placed on the plate. However, after a short period of time the smoke died off, and there was no additional indication of reaction.

As a test, when it was apparent that no further visual signs of reaction were present, the wind was turned off with the fuel bed still on the plate during one of the tests around 350 °C. The fuel began to smoke immediately, and flames appeared quickly when the wind was turned back on after two minutes. These observations indicate that for plate temperatures around 350 °C the high wind within the fuel bed convectively cooled the fuel to a point where it could not sustain glowing combustion.

As the plate temperature was lowered from around 350 °C to around 340 °C, a similar behavior was seen with low wind flows, with only one of the three tests igniting at the lower plate temperature. As for the high-wind cases around 350 °C, the fuel bed began to smoke again when the wind was turned off. In one case, glowing combustion was observed when the wind was turned back on after a couple of minutes.

Shredded newsprint placed on the heated plate at 333 °C and exposed to the low wind also failed to develop glowing combustion, but light smoke did appear shortly after the fuel was applied and then dissipated and died off. Figure 25 shows photographs of the bottom and top of this fuel bed after removal from the plate. Inspection shows that the shredded newsprint was heavily blackened on the downwind side of fuel bed base and that there was a light brown band evident across the top of the fuel bed perpendicular to the wind flow about ¾ of the way across the bed from the upstream edge. The absence of blackening on the upstream portion of the fuel bed base provides additional support that the air flow cooled the fuel sufficiently that pyrolysis could not take place, while further downstream the cooling prevented the development of smoldering from the pyrolysis region in immediate contact with the heated plate. The brown band on the top is located at the location where smoke was observed coming from the fuel bed. It may be due to either deposited smoke coming from below or light pyrolysis of the fuel by the heated wind-driven plume passing through the fuel bed.

At temperatures around 340 °C glowing combustion was observed for all three experiments run without an applied wind. There was considerable scatter in the times required for glowing to appear. When the temperature was lowered an additional 10 °C glowing combustion did not develop, even though some smoke was released and blackening of the fuel bed base occurred. This blackening is evident in Figure 26, which shows the bottom and top of the shredded newsprint after removal from the plate held at 332 °C. The blackened newsprint extends across the bottom of the fuel bed except for the outer edges. On the top there is a roughly circular area that is browned. The degree of darkening and the diameter of the circle were found to increase

Figure 25. Bottom (left) and top (right) views of a shredded newsprint fuel bed are shown after the fuel was removed from the heated plate held at 333 °C with a low wind applied. Glowing combustion was not observed.

Figure 26. Bottom (left) and top (right) views of a shredded newsprint fuel bed are shown after the fuel was removed from the heated plate held at 332 °C with no wind applied. Glowing combustion was not observed.

with fuel depth. Similar to the low wind case, this behavior suggests that the browning resulted from soot deposition from or light pyrolysis due to the thermal plume within the porous fuel bed. These images reveal that even though the plate surface temperature was high enough to cause some pyrolysis of the fuel in immediate contact, it was insufficient to allow self-sustained smoldering to develop and did not provide sufficient heat to ignite glowing combustion.

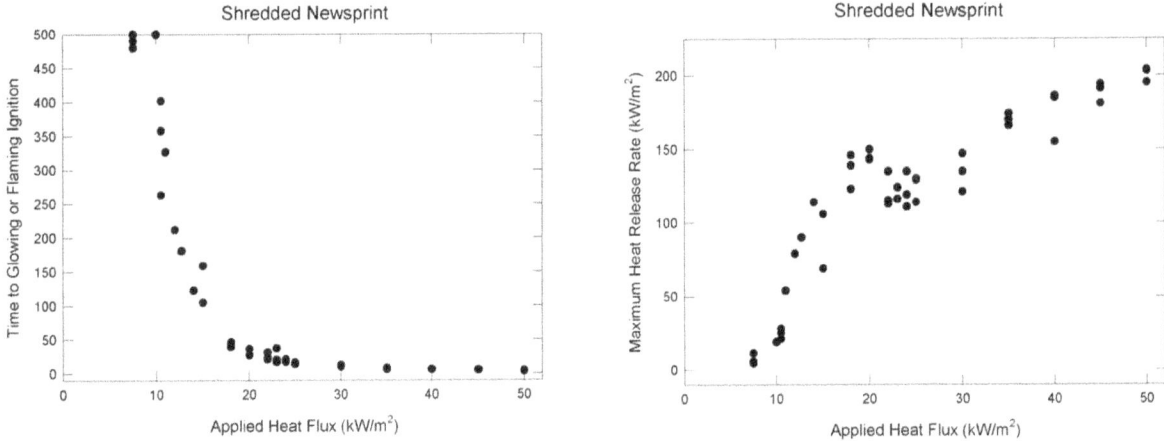

Figure 27. Values of time to glowing or flaming ignition and maximum observed heat release rate are plotted as functions of applied heat flux for shredded newsprint fuel beds.

4.2.2. Radiative Heating Ignition Results

The ignition behaviors of shredded newsprint fuel beds subjected to various applied heat flux (AHF) levels of thermal radiation were characterized using the cone calorimeter. The most direct indicator of ignition behavior dependence on AHF is the time required for flaming ignition. Figure 27 summarizes the time to glowing or flaming observations for shredded newsprint. Neither glowing nor flames appeared with very low AHFs. These experiments are indicated by the values plotted near the upper AHF axis, actual experimental times were usually longer. The heat release rate (HRR) behavior is widely considered the most important fuel flammability parameter. Maximum observed HRRs are plotted as a function of AHF in Figure 27.

For AHFs ranging from 20 kW/m^2 to 50 kW/m^2 times to ignition decreased slowly from about 20 s to around 4 s with increasing AHF. For AHFs below 20 kW/m^2 ignition times began to rise rapidly with decreasing AHF, approaching 350 s for 11 kW/m^2. Flaming ignition did not occur for AHFs around 10 kW/m^2, but glowing ignition was observed, requiring between 250 s and 400 s. Glowing combustion was not observed when the AHF was decreased to 7.5 kW/m^2.

Maximum HRRs showed a comparable dependence on AHF. For AHFs between 20 kW/m^2 and 50 kW/m^2 maximum HRRs increased slowly with increasing AHF. This is the expected behavior and is commonly observed in cone calorimeter experiments since the rate of fuel pyrolysis is expected to decrease at lower AHFs. For AHFs below 20 kW/m^2 the maximum HRRs dropped faster, approaching the noise floor of the experiment for an AHF of 7.5 kW/m^2.

Additional details concerning the response of shredded newsprint to a radiative heat flux can be obtained by considering the time behaviors of the HRR and mass loss behavior. Figure 28 shows plots of these variables as a function of time for three experiments with AHFs of 50 kW/m^2. The results show that the HRRs began to increase immediately, reaching maximum values in 16 s, while the samples mass loss began with the application of the heat flux. These observations are

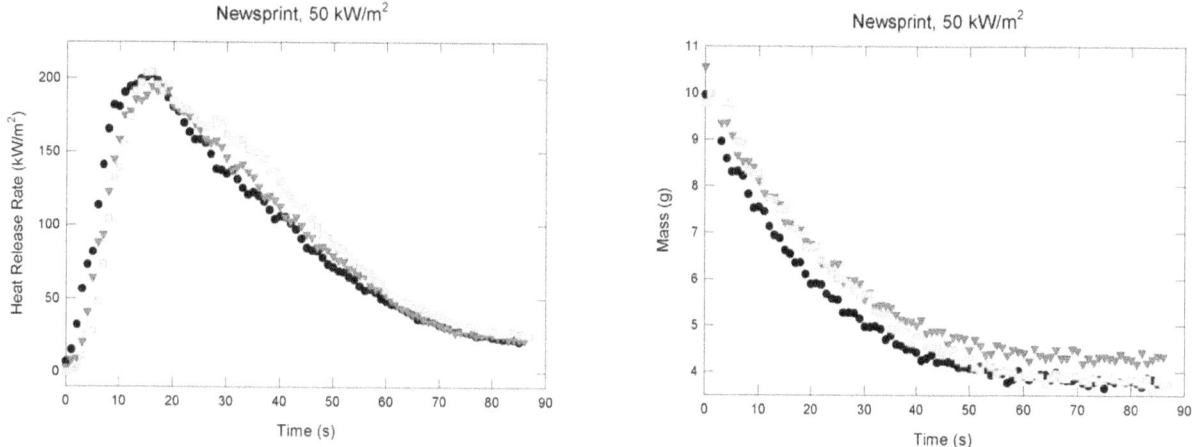

Figure 28. Heat release rates and sample masses are plotted as a function of time for three experiments with a 50 kW/m² heat flux applied to shredded newsprint fuel beds.

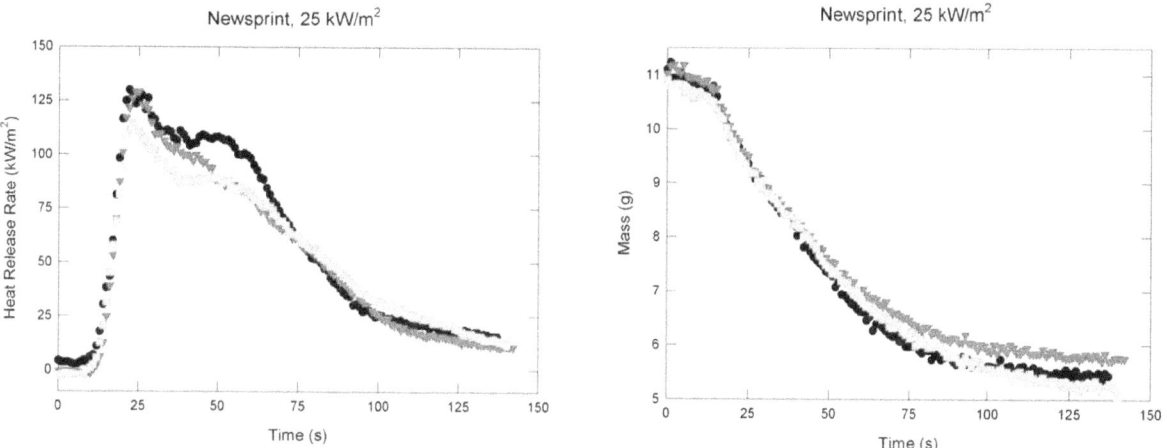

Figure 29. Heat release rates and sample masses are plotted as a function of time for three experiments with a 25 kW/m² heat flux applied to shredded newsprint fuel beds.

consistent with rapid flaming observed with this AHF. Most of the samples HRR and mass loss were complete after 90 s.

Three similar plots are shown in Figure 29 for an AHF of 25 kW/m². The major difference between Figure 28 and Figure 29 is the presence of substantial induction periods lasting approximately 11 s for the 25 kW/m² AHF cases during which there were low mass loss rates and extremely low HRRs. The absence of a measurable HRR during the induction periods indicates that oxidation and heat generation were not occurring even though fuel moisture removal or slow non oxidative pyrolysis, indicated by the mass loss, was taking place. Once oxidation and heat release began, as evidenced by a measurable HRR, the HRR grew rapidly, and flaming was observed within a short period of time, similar to the behavior observed with the 50 kW/m² AHFs. This suggests that once smoldering, and likely glowing combustion, developed there was a rapid transition to flaming.

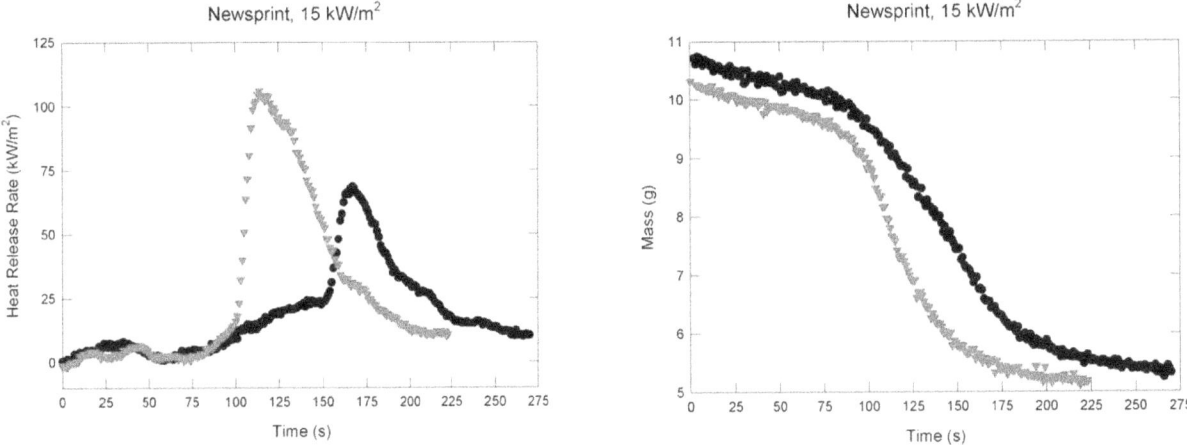

Figure 30. Heat release rates and sample masses are plotted as a function of time for two experiments in which a 15 kW/m^2 heat flux was applied to shredded newsprint.

Figure 30 shows similar plots of HRRs and sample masses for two experiments with an AHF of 15 kW/m^2. The observed data differ markedly from those with higher AHF values described above. As for an AHF of 25 kW/m^2, there is an induction period following the application of the heat flux during which mass loss is observed, but there is no measurable HRR. For 15 kW/m^2 this period lasted about 75 s. After this time the HRR began to increase slowly for both experiments, and there is a marked increase in the mass loss rate. The relatively small values and initial slow growth of the HRRs suggest that smoldering had developed and was spreading within the fuel beds.

The two experiments had different behaviors once smoldering developed. For one, the HRR seemed to grow more quickly, and there was a distinct transition to flaming roughly 25 s after smoldering developed that resulted in a sudden large increase in HRR by nearly 100 kW/m^2, with a substantial increase in the mass loss rate at the same time. For the second experiment the smoldering phase lasted much longer (roughly 75 s), and the HRR due to smoldering was much higher. When flaming did develop around 150 s, the HRR increase and overall peak values were considerably reduced compared to the other case. Apparently, the longer smoldering time resulted in a reduction in the rate of fuel pyrolysis during flaming as well as the fuel available to support flaming. The mass loss behavior also reflects this difference, and the mass loss rate was lower during flaming for the second experiment.

For AHFs as low as 11 kW/m^2 the HRR and mass behaviors were similar to those in Figure 30. For 11 kW/m^2 there was an induction period that lasted roughly 200 s, followed by a period of smoldering during which the HRR and mass loss slowly increased with time, followed by a short flaming period that appeared roughly 125 s after smoldering first started. It is the sum of the induction period followed by the smoldering period that combine to give the observed flaming ignition times.

As seen in Figure 27, flaming ignition was not observed for AHFs of 10.5 kW/m^2 or less. HRR measurements with AHF around 10 kW/m^2 did show that there was still an induction period during which mass loss took place, but there was no measurable HRR. The induction period was

38

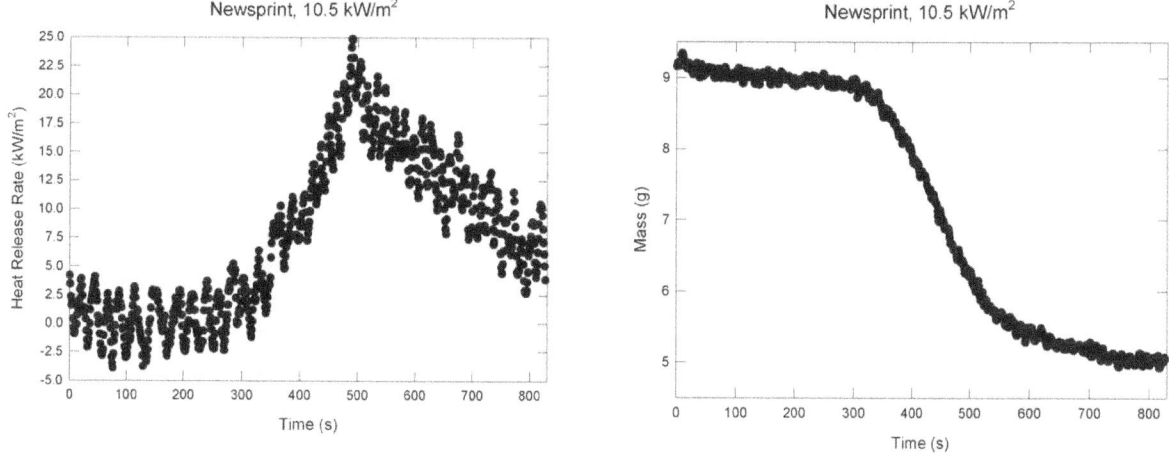

Figure 31. Heat release rates and sample masses are plotted as a function of time for an experiment in which a 10.5 kW/m² heat flux was applied to shredded newsprint.

followed by a period during which smoldering developed and was marked by an initially increasing HRR and accelerated mass loss rate. Figure 31 shows an example of this behavior with an AHF of 10.5 kW/m². Note that nearly 300 s was required for smoldering to develop.

When an AHF of 7.5 kW/m² was used there was a very small amount of mass loss immediately following the application of the heat flux. After this, the mass no longer changed. The HRR remained very close to zero over a 600 s period.

Figure 32 shows the appearance of a newsprint fuel bed after applying 7.5 kW/m² for several minutes. There is some light browning on the top of the bed, but no indication of a self-sustained reaction. The non uniform browning may be associated with small spatial variations in the heat flux from the cone heater or to variations in cooling associated with natural flow from the heated fuel surface.

4.3. May Tall Fescue

Both heated plate and cone ignition experiments were done for this fuel.

4.3.1.   Heated Surface Ignition Results

Figure 33 summarizes the temperature dependence of the times required for glowing combustion and flaming following the application of May tall fescue fuel beds to the heated plate. The most striking feature of the data is the nearly complete absence of flaming combustion. For the 28 cases in which glowing combustion was observed, only three resulted in flaming. It should be noted that even though a transition to flaming did not occur, the glowing combustion could be quite intense, particularly with an applied wind. When flaming did appear, it was weak and short lived. There is no apparent temperature dependence for whether flaming occurred or not. The flaming behavior for May tall fescue contrasts with that found for shredded newsprint, where

Figure 32. Photograph of the top surface of a shredded newsprint fuel bed following application of a 7.5 kW/m^2 heat flux.

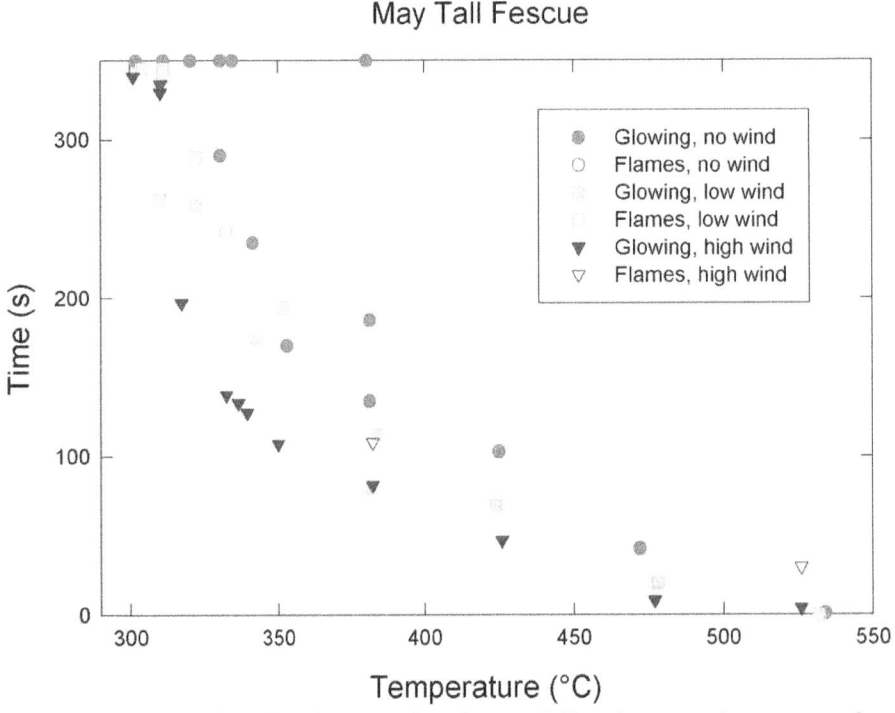

Figure 33. Ignition times for glowing combustion and flaming are shown as a function of temperature for May tall fescue grass fuel beds applied to a heated plate. Results are included for no wind, low wind, and high wind cases.

flaming combustion took place in nearly every experiment in which glowing combustion was observed.

Even though there is scatter in the glowing combustion data, it is clear that the times required for ignition fall on three distinct bands corresponding to no wind, low wind, and high wind conditions. The times required for ignition increased from near zero around 525 °C to periods approaching 300 s for temperatures around 325 °C. Similar to observations for the shredded newsprint, the ignition times for a given temperature decreased with increasing wind. Unlike for the shredded newsprint, the largest reductions occurred on going from low wind cases to high wind cases.

The proportional change in going from no wind to low wind seems smaller than for the shredded newsprint. At the lower temperatures the reduction in ignition times on going from no wind to high wind appears to be greater than a factor of two for the grass.

There are other differences between the May tall fescue and shredded newsprint results. Even though the shredded newsprint transitions to flaming much more readily than the grass, the May tall fescue seems to smolder more easily. This can be seen by comparing times required for glowing combustion at a given temperature, which are generally lower for the grass, and noting that glowing combustion develops for lower plate temperatures (as low as 310 °C) than for the shredded newsprint (as low as 340 °C).

Another difference between the shredded newsprint and May tall fescue concerns the lowest temperatures for which smoldering or flaming was observed for the three wind conditions. For newsprint, combustion appeared at lower plate temperatures as the wind was decreased, with the lowest temperatures capable of igniting the newsprint being observed for the no-wind condition. The opposite is true for the data shown in Figure 33, with the lowest ignition temperatures observed with wind present. This observation suggests that for the low temperature end of the surface ignition curve the development of glowing combustion for May tall fescue is more sensitive to the amount of air supplied to the fuel surface than is shredded newsprint, since the additional air effectively counteracts the additional cooling resulting from the air flow.

The appearance of the fuel beds following an experiment provides additional insights into the importance of air flow on the reaction behavior of the May tall fescue fuel beds. Figure 34 shows bottom and top views of a fuel bed that had developed glowing combustion after being placed on the heated plate held at 341 °C without an applied wind. It is clear that the smoldering spread through the entire bed. There is a narrow region of unburned fuel around the three edges of the fuel bed with wire-screen extending below the plate. No unburned fuel is evident on the side of the cage with the narrow open slit at the base of the wire-screen side wall. In fact, the remaining fuel at this location seemed to be grayer, suggesting more complete combustion. These observations agree with the earlier discussion that the air inflows on the cage side wall with the open slit differs from that through the sides with full wire-screen sidewalls.

Figure 35 shows photographs of the bottom and top of a fuel bed following removal from the heated plate held at 336 °C with a high wind applied. Glowing combustion was observed at

Figure 34. Bottom (left) and top (right) views of a May tall fescue fuel bed are shown after the fuel was removed from the heated plate held at 341 °C with no wind applied. Glowing combustion was observed after 235 s.

Figure 35. Bottom (left) and top (right) views of a May tall fescue fuel bed are shown after the fuel was removed from the heated plate held at 336 °C with a high wind applied. Glowing combustion was observed after 134 s.

134 s following application to the plate and was allowed to proceed for a period of time. The effects of the air flow are marked. The fuel bed has three distinct bands that run perpendicular to

Figure 36. Bottom (left) and top (right) views of a May tall fescue fuel bed are shown after the fuel was removed from the heated plate held at 311 °C with no wind applied. Glowing combustion was not observed.

the wind direction and extend from the top to the bottom of the bed. There is an area of unburned grass on the upstream side of the bed. Apparently, cooling associated with the air flow was sufficient to prevent propagation of the glowing combustion into this area. Near the center there is a grayish area indicating more complete combustion. This is likely the result of more intense reaction due to a greater availability of air. On the downstream side of the fuel bed the tall fescue is simply blackened, suggesting incomplete reaction of the fuel. This type of burning pattern is typical of those observed with an applied wind, but with the low wind the unburned upstream portion was narrower, and the gray band was located closer to the upstream edge.

It was noted above that glowing combustion for the May tall fescue was not observed below 340 °C in the absence of a wind. However, the appearance of the fuel beds following exposure to the heated plate at lower temperatures indicates that non glowing smoldering was still taking place at much lower plate temperatures. Figure 36 shows the bottom and top of a fuel bed that was held at 311 °C without an applied wind. The bottom of the fuel bed is heavily blackened across its entire extent. On the top there is circle of similar blackened material surrounded by a band of unburned grass. Such blackening is not expected in the absence of fuel surface oxidation. These observations indicate that a non glowing smolder wave started near the heated plate and passed upward to the top surface of the fuel bed. The most likely reason for the area of unburned grass around the outside of the top surface is that this grass was cooled by air being entrained in the thermal plume rising from the heated surface and smoldering fuel.

For the lowest plate temperature tested without an applied wind, 302 °C, the fuel bed had a similar appearance to that seen in Figure 36.

Figure 37.   Bottom (left) and top (right) views of a May tall fescue fuel bed are shown after the fuel was removed from the heated plate held at 311 °C with a low wind applied. Glowing combustion was not observed.

Figure 37 shows similar photographs for a fuel bed held at the same surface temperature, 311 °C, as in Figure 36, but with a low wind applied. Most of the bottom is blackened with the exception of a narrow band along the upwind (right hand side in photograph) edge. On the top surface of the fuel bed there is a brown band that extends across the bed perpendicular to the wind direction and just downstream of the center. Deeper within the bed the band is much darker. Clearly there had been some pyrolysis of the fuel in contact with the plate and in the vicinity of the line plume formed by heated gases rising within the fuel bed. Smoldering does not seem to have occurred, since there is no indication that areas of pyrolysis moved away from the heated areas.

Tests with low and high winds with plate temperatures near 300 °C showed partial blackening of the bottoms of the fuel bed, but there were larger areas of unblackened fuels along the upstream edges. For the high wind case, this area extended nearly half way across the bed. For both low and high wind cases there were light brown bands across the tops of the fuel bed, with the band for the high wind case located further downstream.

4.3.2.   Radiative Heating Ignition Results

The heated plate experiments for May tall fescue showed that these fuel beds were unlikely to transition to flaming, even when glowing combustion was present. With one exception, flaming was only observed in the radiative ignition experiments for AHFs equal to or greater than 40 kW/m^2. On the other hand, glowing combustion was observed with much lower AHFs. Figure 38 shows the times required for glowing or flaming combustion to appear following application of the radiative heat flux as a function of the AHF. Similar to the results for shredded newsprint (see Figure 27), the ignition times increased very slowly as the AHF was reduced from

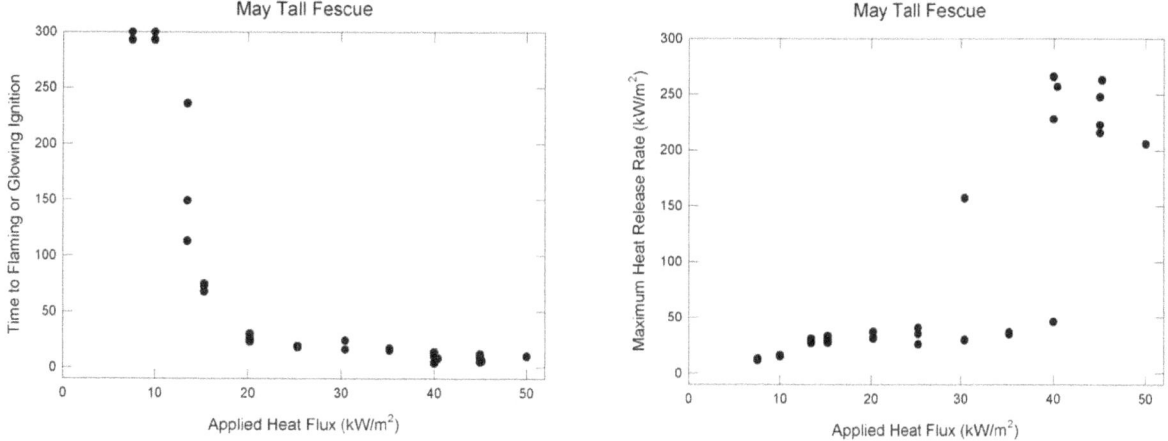

Figure 38. Values of time to flaming or glowing ignition and maximum observed heat release rate are plotted as functions of applied heat flux for May tall fescue fuel beds.

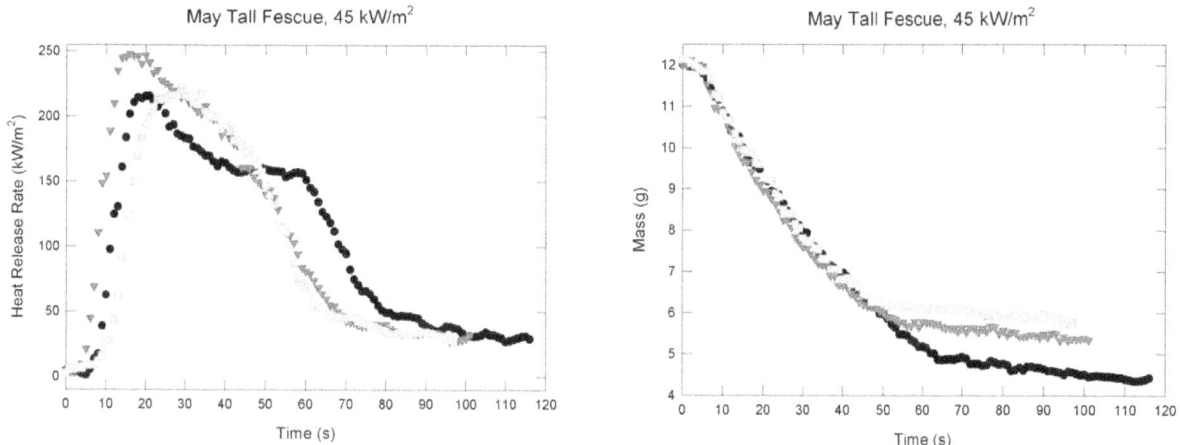

Figure 39. Heat release rates and sample masses are plotted as a function of time for three experiments in which a 45 kW/m^2 heat flux was applied to May tall fescue.

50 kW/m^2 to 20 kW/m^2. For still lower AHFs the ignition times began to increase rapidly with decreasing AHF before ignition was no longer observed with AHFs near 10 kW/m^2 and lower.

Even though the ignition time plots for shredded newsprint and May tall fescue have similar appearances, distinct differences in burning behavior appear in the plots of maximum HRR versus AHF. Maximum HRRs for the shredded newsprint remained high for AHF as low as 11 kW/m^2, while maximum HRRs for the May tall fescue dropped to low values for AHFs lower than 40 kW/m^2. The abrupt drop in HRR for the May tall fescue around an AHF of 40 kW/m^2 is due to the absence of flaming combustion at lower AHFs.

Plots of HRR and fuel mass versus time provide additional insights into the ignition behaviors of the May tall fescue fuel beds. Figure 39 shows three sets of results for an AHF of 45 kW/m^2. After a brief period during which there was a relatively slow loss of fuel mass and low HRR, both the HRRs and mass loss rates increased rapidly, reflecting the onset of flaming. Flaming

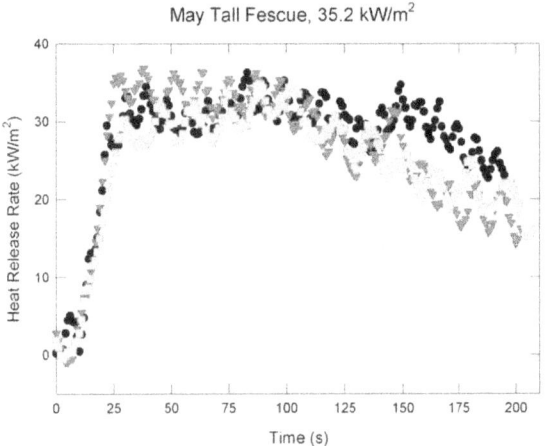
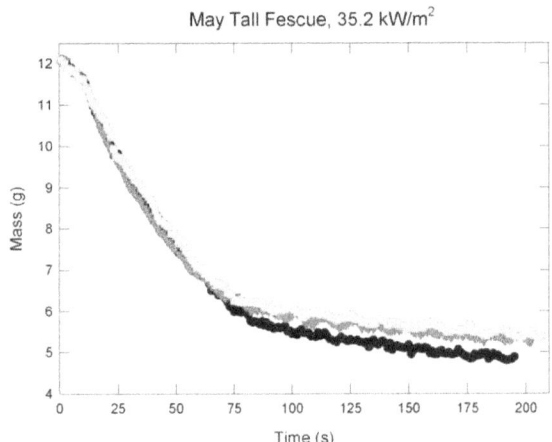

Figure 40. Heat release rates and sample masses are plotted as a function of time for three experiments in which a 35.2 kW/m^2 heat flux was applied to May tall fescue.

lasted for about 50 s to 60 s before dying down. Apparently, smoldering combustion continued after the flaming period since measurable HRRs and low mass lost rates were still present.

Flaming was not observed when an AHF of 35.2 kW/m^2 was applied to the May tall fescue. Figure 40 shows HRRs and masses for three experiments. Similar to Figure 39, there is a brief induction period lasting about 10 s during which there is little HRR, but mass loss is taking place. After this time, the HRRs began to increase rapidly, rising to values on the order of 32 kW/m^2. HRR values remained nearly constant for about 50 s before beginning to fall slowly. The fuel mass loss rate abruptly accelerated at the same time the HRR first begins to rise. This is near the time when glowing combustion first appeared, suggesting that glowing combustion spreading over the fuel bed is responsible for the increased HRR and mass loss rate. At roughly the same time that the HRR began to drop slowly there was a distinct change in the slope of the mass loss, with the mass loss rate decreasing abruptly.

When the effective heat of combustion (EHC) is defined as the HRR divided by the mass loss rate, it is clear that the EHC abruptly increased at the time when the mass loss rate drops. These observations suggest that at least two types of fuel surface oxidation reactions were taking place. The first has a relatively low EHC and likely corresponds to oxidative surface reaction of easily pyrolysed fuel components. Initial pyrolyis of cellulosic fuels is known to form an enriched-carbon char that has a higher EHC that reacts more slowly. The oxidation of the char formed during the initial more rapid oxidative pyrolysis of the grass is most likely responsible for the period of relatively high HRR with relatively low mass loss rate.

In Figure 39 it is evident that the HRR and mass loss behavior following the flaming periods are similar to those observed at the longest times in Figure 40. This suggests that the flaming observed for the May tall fescue with the highest AHFs also formed a high energy containing char that then oxidized more slowly.

Comparison of the HRR and mass loss behaviors observed for AHFs over the range of 20 kW/m^2 to 35 kW/m^2 showed that they were similar to those in Figure 40 with some variation in slopes

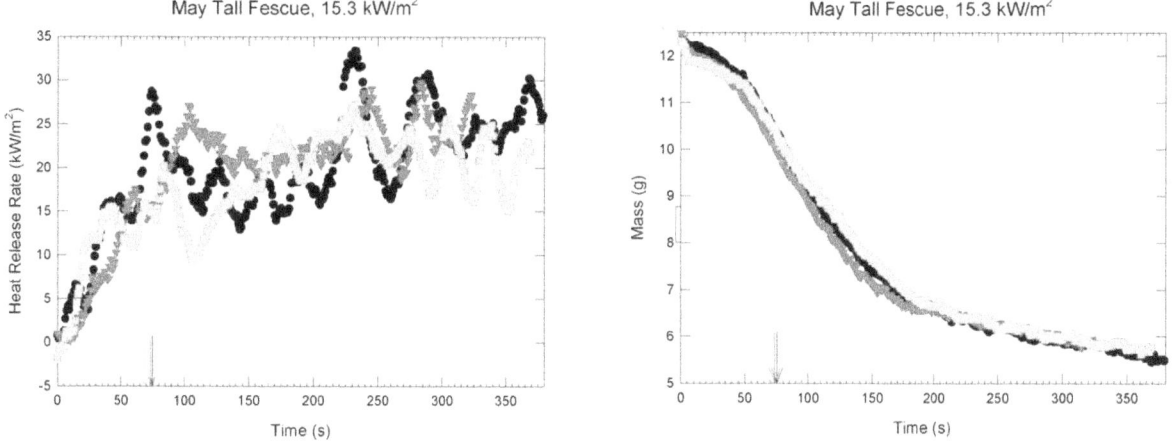

Figure 41. Heat release rates and sample masses are plotted as a function of time for three experiments in which a 15.3 kW/m^2 heat flux was applied to May tall fescue. The arrows indicate the times when glowing combustion appeared.

and transition times. The relatively short times required for glowing combustion to develop for these AHFs (see Figure 38) suggest that glowing combustion rapidly developed and spread over the fuel bed, leading to the formation of chars that then oxidized more slowly.

For AHFs less than 20 kW/m^2 the time required for glowing combustion to develop rapidly increased. Figure 41 shows three sets of measured HRRs and masses as a function of time for an AHF of 15.3 kW/m^2. Color-coded arrows have been added to the plots indicating the times when glowing combustion was observed. The data show that low HRRs developed shortly after the fuel was exposed to the AHF. Relatively slow mass loss rates appeared around the same time. Around 50 s after exposure there was a clear shift in the mass behavior, with the mass loss rate increasing. This change is most likely associated with the development of sustained smoldering. Interestingly, glowing combustion did not appear until about 20 s after smoldering developed. This suggests that the initial smoldering is non glowing and that transition to glowing required a short period of time. These conclusions are consistent with findings from the heated plate experiments that showed that non glowing smoldering occurred at lower temperatures.

The observation of nearly constant HRRs while there is a distinct reduction in the mass loss rate around 175 s indicates that two distinct EHCs occurred for this lower AHF. This behavior is similar to that observed with AHFs covering the 20 kW/m^2 to 35 kW/m^2 AHF range.

Figure 42 shows similar plots for three experiments with an AHF of 13.5 kW/m^2. For each there was a short induction period at the start of the heating during which the HRR remained close to zero but during which there was a slow mass loss. After the induction period the HRR began to rise, but the increases were much slower than when 15.3 kW/m^2 was used. At roughly 75 s there was a distinct increase in the mass loss rate suggesting that smoldering had developed. Glowing combustion was not observed until much later when a significant fraction of the more rapid mass change had already occurred. The sustained HRR and reduced mass loss rate at longer times again indicates that an initial pyrolysis produced a char that then oxidized with a higher effective

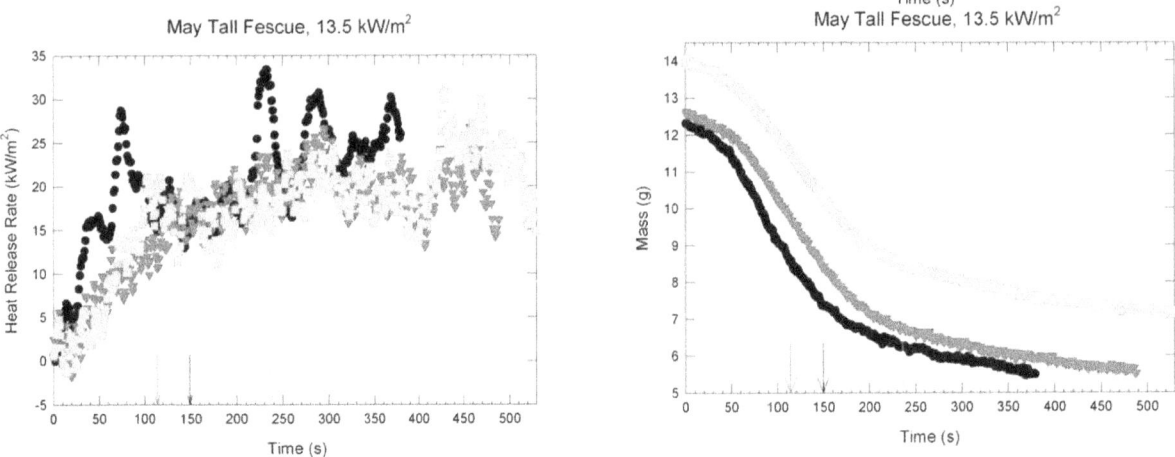

Figure 42. Heat release rates and sample masses are plotted as a function of time for three experiments in which a 13.5 kW/m² heat flux was applied to May tall fescue. The arrows indicate the times when glowing combustion appeared.

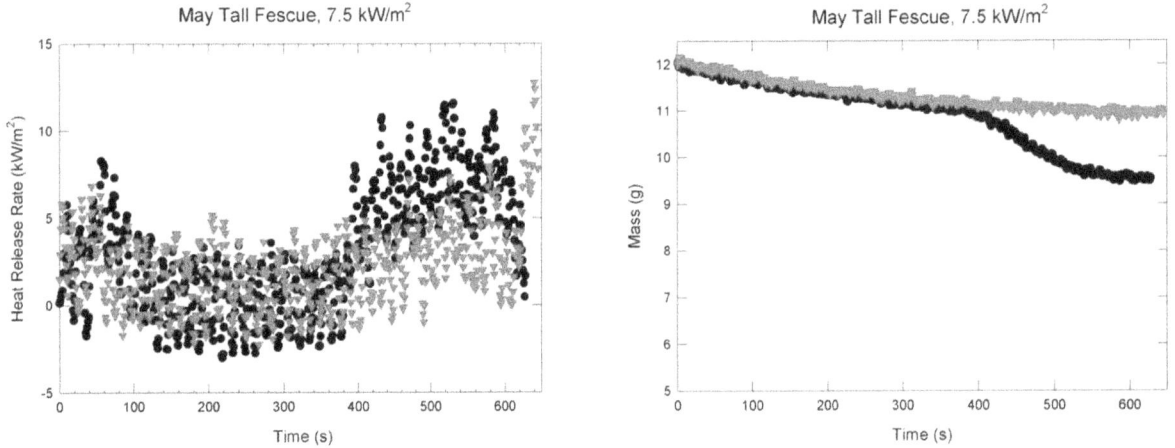

Figure 43. Heat release rates and sample masses are plotted as a function of time for two experiments in which a 7.5 kW/m² heat flux was applied to May tall fescue.

heat of combustion. The much longer periods required for each of these stages as compared to cases with AHFs of 15.3 kW/m² requires a strong dependence on AHF.

Glowing combustion was not observed for experiments with AHFs of 10 kW/m² and 7.5 kW/m². Figure 43 shows plots of HRR and mass as functions of time for two experiments with an AHF of 7.5 kW/m². After the heat fluxes were applied there were long periods lasting several hundred seconds during which a very slow mass loss occured without significant HRR. The mass lost could be due to non oxidative pyrolysis or moisture removal. During one of the experiments the mass loss rate accelerated around 400 s. There seems to have been a very small increase in the measured HRR at the same time, suggesting limited oxidation was taking place. Interestingly, both the HRR and mass loss rate seemed to decrease back to the lower levels after about 100 s, a time when a large fraction of the original mass remained. This suggests that even though some

Figure 44. This slightly out of focus photograph shows the top surface of a May tall fescue fuel bed following application of an applied radiative heat flux that was insufficient to induce glowing combustion.

non glowing oxidation took place, there was no sustained smoldering. There is a hint that the HRR for the second run may have started to increase after 600 s, just as the experiment ended.

Results with AHFs of 10 kW/m^2 were similar to those for 7.5 kW/m^2 with a nearly linear mass change over the first 200 s followed by an increased mass loss rate lasting about 100 s. There seemed to be a very small HRR that only developed at roughly the same time as the change in mass lost rate. After 600 s about 20 % of the original mass remained.

Some blackening of the grass was observed at the lower AHFs. Figure 44 shows an example of a fuel bed at the conclusion of an experiment in which glowing combustion did not occur. Some blackening of the surface is evident. The blackening is not uniform, but has a distribution over the top of the fuel bed that is similar to that seen for the shredded newsprint in Figure 32. As before, this distribution could be due to the air flow distribution or small variations the cone thermal radiation distribution.

4.4. August Tall Fescue

Only heated plate ignition experiments were performed for August tall fescue. Figure 45 shows the measured ignition times for glowing combustion and flaming as a function of the heated plate temperature. The results will be compared with the corresponding results for May tall fescue shown in Figure 33.

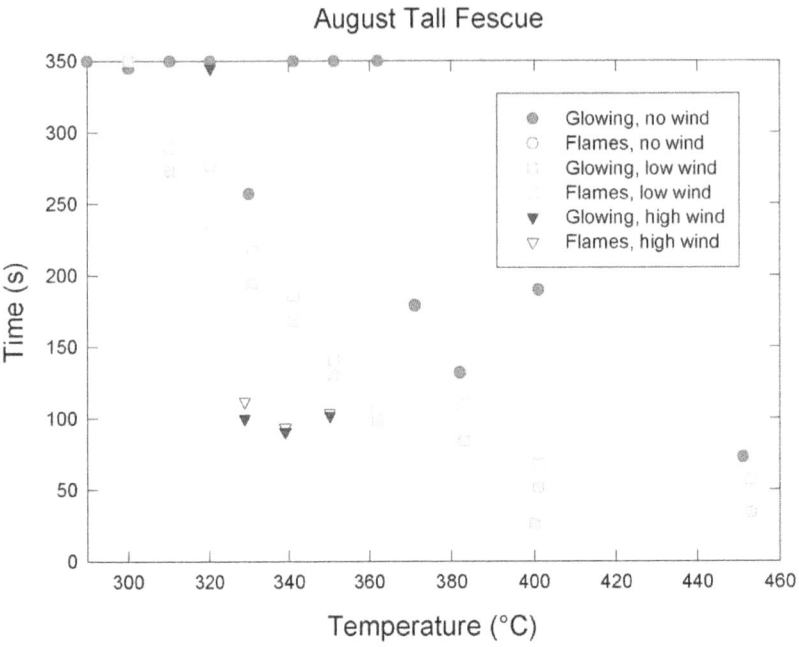

Figure 45. Ignition times for glowing combustion and flaming are shown as a function of temperature for August tall fescue grass fuel beds applied to a heated plate. Results are included for no wind, low wind, and high wind cases.

The most dramatic difference between the two fuels is that the August tall fescue was much more likely to transition to flaming in the presence of a wind. Only three such transitions with limited flaming times were observed with the May tall fescue, while all of the August tall fescue samples exposed to the wind transitioned to flaming. These flames were robust and lasted for many seconds. Figure 46 shows an example of the flames for an experiment with the heated plate held at 310 °C in the presence of the low wind. Flames were observed 289 s after the fuel was applied to the plate and the photograph was taken 5 s later. The strong glowing that tended to develop along the upstream edge with an applied wind is also visible in this image.

The transition to flaming for August tall fescue required the presence of a wind. Flames were not observed for the five experiments without wind in which glowing combustion developed.

Even though flaming was more likely for August tall fescue than for May tall fescue, the August tall fescue required higher heated plate temperatures to induce glowing combustion in the absence of wind. With the one exception of a measurement with the plate temperature held at 380 °C, glowing combustion for the May tall fescue was observed for plate temperatures at or above 340 °C and was absent for plate temperatures around 320 °C. Both responses were observed with the plate temperature held near 330 °C. For the August tall fescue, again with the exception of one experiment, the glowing combustion was observed for temperatures of 370 °C and higher, while it was absent for temperatures of 360 °C and lower, suggesting a 30 °C difference between the two grass samples for the lowest temperatures required to induce glowing combustion when wind was absent.

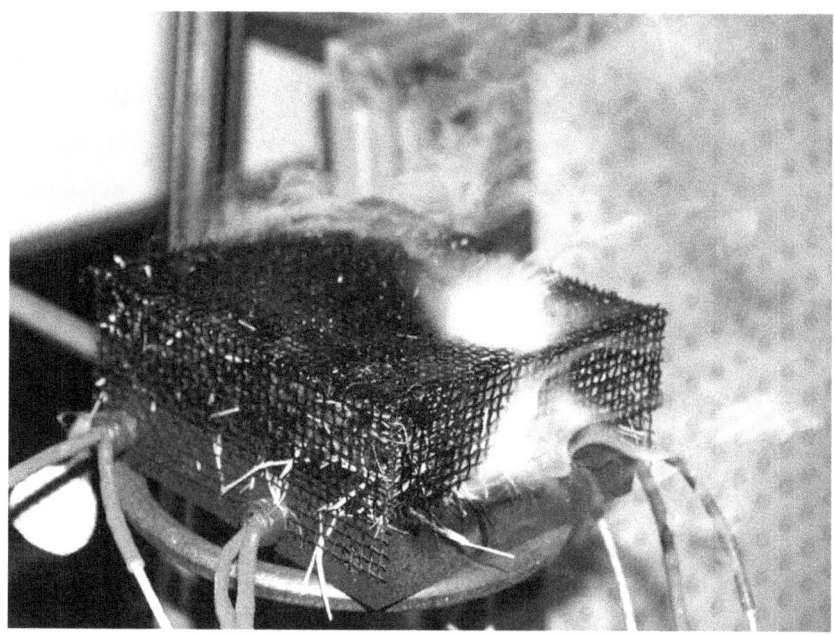

Figure 46. Photograph showing flaming from an August tall fescue fuel bed on the heated plate held at 310 °C in the presence of a low wind. Flames appeared 289 s after the fuel was placed on the plate, and the photograph was taken 5 s later. Strong glowing combustion is present along the upstream edge of the fuel bed.

Despite the fact that the transition to flaming behaviors for the two grasses were very different in the presence of a wind, the dependencies of glowing combustion ignition times on the plate temperature were similar. For both the lowest plate temperature for which glowing combustion developed in the presence of a low wind was around 310 °C, and the times required for ignition were similar, with roughly 275 s being required around 310 °C. The data for the August tall fescue indicate that the glowing combustion ignition times in the presence of a low wind are reduced from those in the absence of the wind, in a manner similar to those for the May tall fescue. Recall that the effect of the low wind was somewhat less than observed for the shredded newsprint.

The high wind results for August tall fescue are limited. However, similar to the May tall fescue, the times required for glowing combustion ignition fall somewhat below those with a low wind. For the May tall fescue glowing combustion, ignition was observed for plate temperatures as low as 317 °C, while the lowest ignition temperature for August tall fescue was 329 °C. Due to the limited measurements, it is not possible to determine if this difference is statistically significant.

The appearance of the fuel beds following experiments in which ignition was not observed provides additional insight in the smoldering of August tall fescue. Even though glowing combustion was not generally observed for temperatures below 360 °C without a wind, images indicate that non glowing smoldering developed and spread partially through the fuel beds with temperatures as low as 300 °C. Figure 47 shows the bottom and top of a fuel bed after being placed on the heated plate at 300 °C. The bottom of the fuel bed is blackened over a large fraction of its area. Three narrow bands of unblackened fuel are visible along the edges. The top

Figure 47. Bottom (left) and top (right) views of an August tall fescue fuel bed are shown after the fuel was removed from the heated plate held at 300 °C with no wind applied. No glowing combustion was observed.

of the fuel bed is also blackened, indicating that smoldering had passed upward through the bed even though the smoldering seems to have progressed upward primarily on the right side of the fuel bed and on the side facing the camera where the open slit was present in the wire-screen cage. At higher plate temperatures where glowing combustion was not observed, the fuel beds had similar appearances, but the fractions of the bottom and top surfaces blackened were larger.

Figure 48 shows the appearance of the bottom and top of an August tall fescue fuel bed after removal from the heated plate held at 290 °C without a wind applied. A blackened area covers much of the bottom of the fuel bed, but on the top the only indication of the heating is a small circular brownish area near the center. As discussed earlier, this browning may be due to the deposit of smoke in the thermal plume coming from below or to light heating of the fuel by the thermal plume. In either case, it is clear that smoldering did not propagate upward to the top of the fuel bed from below. While sufficient to cause pyrolysis of the fuel in direct contact with the heated plate, this plate temperature was too low to generate sustained smoldering.

A fuel bed that was placed on the heated plate held at 310 °C with a low wind applied is shown in Figure 49. This is the same fuel bed shown in Figure 46. Both glowing combustion and flaming were observed. Much of the fuel bed is blackened, but there is an area of unburned grass on the downstream edge of the bed. This suggests that air entrained into the burning fuel bed was able to cool the fuel sufficiently to limit glowing combustion and flame spread. In the image showing the bottom of the fuel bed there is a small gray area on the upwind edge (right side of photograph) indicating that some ash formation had occurred. This suggests that complete oxidation of the char formed by the initial fuel pyrolysis was aided by the presence of the wind.

Figure 48. Bottom (left) and top (right) views of an August tall fescue fuel bed are shown after the fuel was removed from the heated plate held at 290 °C with no wind applied. No glowing combustion was observed.

Figure 49. Bottom (left) and top (right) views of an August tall fescue fuel bed are shown after the fuel was removed from the heated plate held at 310 °C with a low wind applied. Glowing combustion and flaming were observed.

The appearance of fuel beds that did not develop glowing combustion in the presence of a wind provides evidence for the important role that convective cooling played in hindering the development of smoldering at low plate temperatures. Figure 50 shows the bottom and top of the fuel bed following application to the heated plate held at 320 °C in the presence of a high wind. On the bottom there is a band of blackened fuel perpendicular to the wind direction that

Figure 50. Bottom (left) and top (right) views of an August tall fescue fuel bed are shown after the fuel was removed from the heated plate held at 320 °C with a high wind applied. No glowing combustion was observed.

covers about ¼ of the bed width on the downstream side starting near the center of the bed. This distribution suggests that the wind passing through the upstream side of the cage as well as the air entrained into the resulting thermal plume from the downstream side were effective in cooling the fuel to a level where significant pyrolysis could not take place. The top of the fuel bed shows only a very narrow band of browning due either to smoke deposition or light pyrolysis. These photographs show that significant smoldering did not develop within the fuel bed.

The fuel beds for the cases with low wind for which glowing combustion was not observed had similar appearances to those for the high wind case shown in Figure 50. The area subject to blackening on the bottoms of the fuel surfaces and the degree of browning at the top varied with the heated plate temperature.

4.5. Cheat Grass

Both heated plate and cone calorimeter ignition experiments were done for this fuel.

4.5.1. Heated Surface Ignition Results

Figure 51 shows a plot of times to glowing combustion and flaming as a function of heated plate temperature for cheat grass fuel beds. As observed for the other fuels discussed thus far, even though there is scatter in the experimental results, the data fall on three broad bands corresponding to no wind, low wind, and high wind cases, with the ignition times increasing as the plate temperature decreases. Considering the results for glowing combustion, when ignition occurs for a given plate temperature, the data for no wind lies at longer times, followed by the low-wind results, with the shortest times required for the high-wind cases.

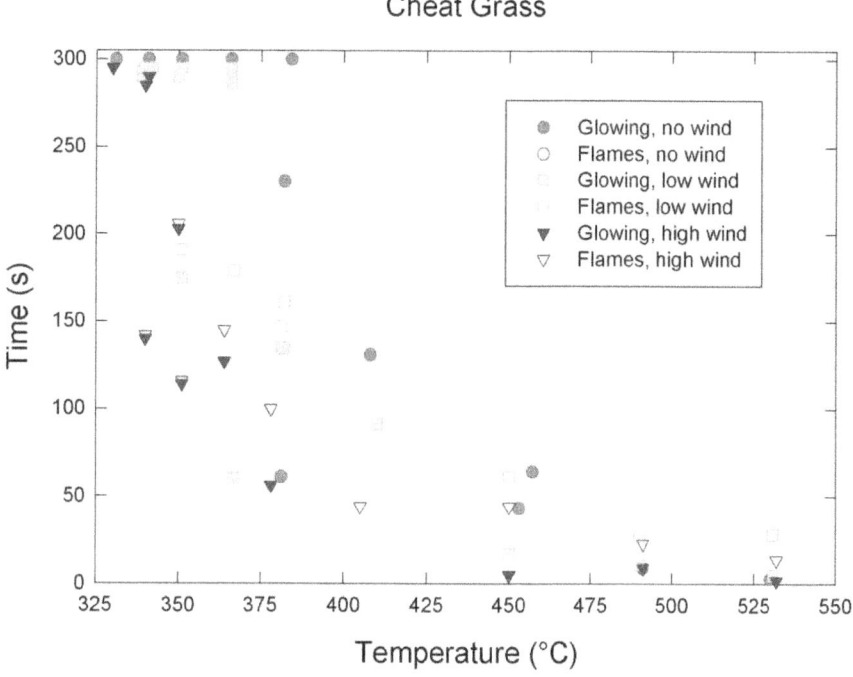

Figure 51.   Ignition times for glowing combustion and flaming are shown as a function of temperature for cheat grass fuel beds applied to a heated plate. Results are included for no wind, low wind, and high wind cases.

The general behavior of the cheat grass with regard to transition to flaming was the same as observed for the August tall fescue. Flaming was not observed when a wind was not applied to the fuel bed, and transition to flaming occurred for all experiments in which glowing combustion developed in the presence of low or high wind.

More detailed comparison with the results for August tall fescue suggests that cheat grass may have a slightly reduced tendency to develop glowing combustion. For both fuels the results suggest that glowing combustion no longer developed when the plate temperature was reduced from around 380 °C to around 360 °C. However, the lowest temperatures for which glowing combustion were observed for the cheat grass were around 340 °C and 350 °C for low and high winds cases, respectively. The corresponding values for August tall fescue were 310 °C and 330 °C. Recall that these temperatures for the August tall fescue were already higher than observed for the May tall fescue, suggesting the tendency for cheat grass to develop glowing combustion is considerably lower than for the May tall fescue.

The appearance of cheat grass fuel beds for which glowing combustion did not develop in the absence of a wind provides additional evidence that the tendency for the cheat grass fuel beds to smolder (either non glowing or glowing) was reduced compared to the August tall fescue. Figure 52 shows photographs for a fuel bed that did not develop glowing combustion when placed on the heated plate held at 341 °C. The bottom of the fuel bed is heavily blackened indicating significant pyrolysis took place. On the top of the fuel there is a darkened circular area near the center that indicates that non glowing smoldering spread upward through the fuel bed. This darkened area is surrounded by fuel that is not discolored and the darkened area is not

Figure 52. Bottom (left) and top (right) views of cheat grass fuel bed are shown after the fuel was removed from the heated plate held at 341 °C with no wind applied. No glowing combustion was observed.

as blackened as the bottom of the fuel bed, suggesting that smoldering was barely sustained. Similar fuel bed photographs for the heated plate held at 331 °C show that even though the fuel was pyrolyzed on the bottom, smoldering did not reach the top surface, which was barely browned near the center. For August tall fescue, non glowing smoldering was observed for plate temperatures down to 300 °C, but not for 290 °C.

The appearance of the cheat grass fuel beds that did not develop glowing combustion in the presence of low and high winds was similar to those already described for other fuels, with blackening on the bottom primarily on the downstream side and a narrow brown band across the top perpendicular to the wind flow direction. Figure 53 shows an example of this for a heated plate temperature of 350 °C in the presence of a low wind. Once again the cooling effect of an applied wind has been shown to be important for the fuel reaction behavior at low heated plate temperatures.

4.5.2. Radiative Heating Ignition Results

Figure 54 shows plots of the times required for glowing or flaming ignition and maximum HRRs as functions of AHF for the cheat grass. Similar to the results for May tall fescue (see Figure 38), flaming was only observed for the highest AHFs used, but glowing combustion was seen for AHFs extending down to approximately 10 kW/m^2. The maximum values of HRR measured for flaming cheat grass were roughly 80 % of those measured for the May tall fescue, even though the initial mass of cheat grass was only 11 % less (8.0 g versus 9.0 g). Quantitative comparison shows that slightly more heat was released by the smoldering May tall fescue for given AHFs in the 10 kW/m^2 and 35 kW/m^2 range.

Figure 53. Bottom (left) and top (right) views of a cheat grass fuel bed are shown after the fuel was removed from the heated plate held at 350 °C with a low wind applied. No glowing combustion was observed.

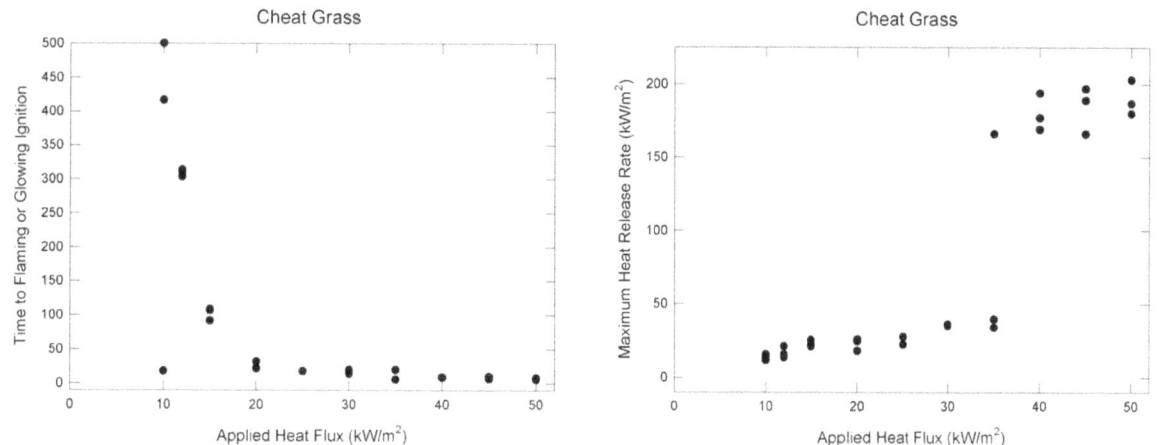

Figure 54. Values of time to flaming ignition and maximum observed heat release rate are plotted as functions of applied heat flux for cheat grass.

The heated plate ignition experiments indicated that cheat grass and May tall fescue had similar behaviors when no wind was applied, but that the grasses displayed very different transition to flaming behaviors when wind was present. Since the cone calorimeter results are similar for the two fuels, it seems likely that radiative heating experiments are not sensitive to fuel behaviors that depend strongly on the presence of wind.

Figure 55 shows the time dependencies of HRR and sample mass measured for an AHF of 45 kW/m^2. Figure 39 shows corresponding results for May tall fescue. The general behaviors are quite similar for the two grasses.

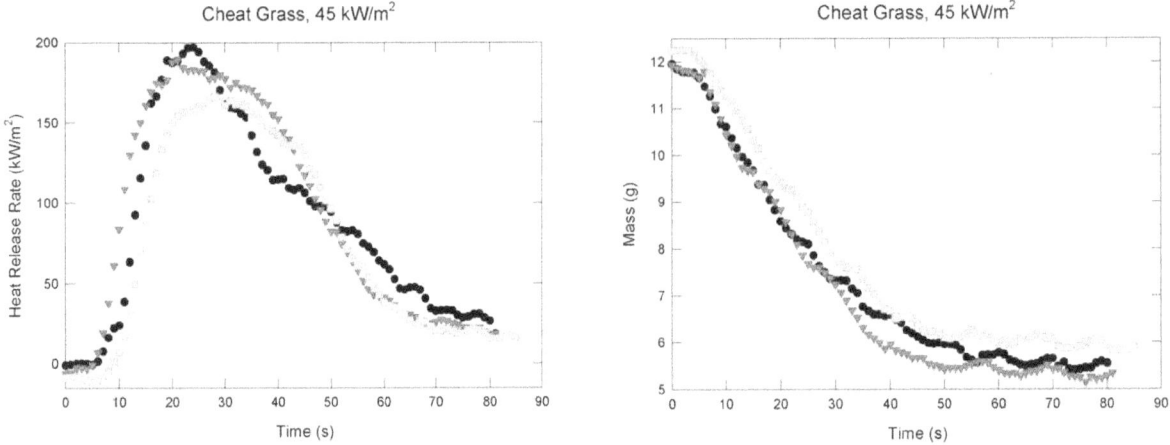

Figure 55. Heat release rates and sample masses are plotted as a function of time for three experiments in which a 45 kW/m^2 heat flux was applied to cheat grass.

Results for HRR and mass loss versus time for experiments with glowing cheat grass had similar time dependences to those for the May tall fescue shown in Figure 40 to Figure 42. Note that this indicates that smoldering cheat grass also has two distinct effective heats of combustion, implying that high energy char was formed and then slowly oxidized.

Glowing combustion was observed for only one of three cheat grass samples run with the AHF set to 10 kW/m^2. All three had HRRs and fuel variations with time similar to those observed for the May tall fescue with an AHF of 7.5 kW/m^2 shown in Figure 43. There were long induction periods lasting about 200 s during which the mass decreased but no heat was released. This was followed by periods during which the mass loss rate was noticeably faster and the HRR rose to about 10 kW/m^2. These observations suggest that non glowing smoldering was taking place at these times. Glowing combustion was observed for one sample at 417 s. By this time the mass loss rate had decreased, indicating that the char formed earlier was being oxidized.

4.6. Fine Florida Grass

Only heated plate ignition experiments were performed for the fine Florida grass. Figure 56 shows the times required for glowing combustion and flaming to appear as functions of heated plate temperature for no, low, and high wind conditions. The results for the different wind cases fall on three distinct curves, with the no wind cases requiring longer times and the high wind cases the shortest times. The reduction in ignition times was largest between the no wind and low wind conditions.

With one exception, flaming was not observed in the absence of wind. In contrast, flaming occurred in each experiment with a low or high wind during which an ignition was noted. In Figure 56 many of the measurements with wind only show the times when flames appeared. For most of these experiments glowing combustion was observed prior to the appearance of flames, but the time differences were too small to be resolved on the plot. The dependence of the transition from glowing combustion to flaming on wind condition is similar to those for August

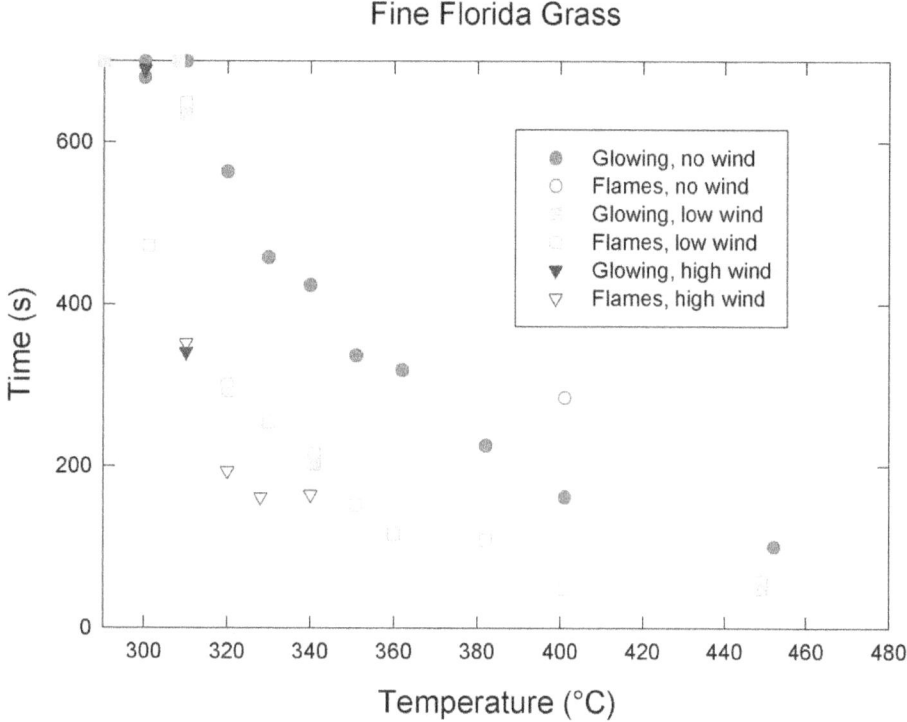

Figure 56. Ignition times for glowing combustion and flaming are shown as a function of temperature for fine Florida grass fuel beds applied to a heated plate. Results are included for no wind, low wind, and high wind cases.

tall fescue (Figure 45) and cheat grass (Figure 51), but the times required for transition are generally shorter. These three types of grass all differ from May tall fescue, since it tended to not flame for any of the wind conditions.

Even though the dependence of ignition behavior on wind condition for the fine Florida grass is the same as observed for August tall fescue and cheat grass, the lowest plate temperatures required for glowing combustion ignition are considerably lower. Ignitions were observed for plate temperatures as low as 321 °C, 301 °C, and 310 °C for the no, low, and high wind cases, respectively. Corresponding values for August tall fescue and cheat grass were 371 °C, 310 °C, and 331 °C and 381 °C, 351 °C, and 340 °C. On the other hand, the lower heated plate temperature ignition limits for May tall fescue (330 °C, 310 °C, and 317 °C, respectively), which had a different transition-to-flaming behavior, are similar to those for the fine Florida grass.

Even though May tall fescue and fine Florida grass have similar lower plate temperature limits for the development of glowing combustion in the absence of wind, comparison of Figure 33 and Figure 56 shows that much longer periods were required for the appearance of glowing at a given plate temperature for the fine Florida grass. This observation, along with the earlier findings that non glowing smoldering developed for August tall fescue and cheat grass for plate temperatures well below those for which glowing combustion occurred, suggests that each of these grasses is capable of non glowing smoldering and that the lower plate temperature limit for glowing

59

Figure 57. Bottom (left) and top (right) views of a fine Florida grass fuel bed are shown after the fuel was removed from the heated plate held at 321 °C with no wind applied. Glowing combustion was observed.

combustion depends on the ease with which non glowing smoldering transitions to glowing combustion.

Figure 57 shows bottom and top views of a fine Florida grass fuel bed after it was removed from the heated plate held at 321 °C. A large fraction of the fuel bed is blackened on the bottom showing that the grass in contact with the heated plate was pyrolyzed. Two areas at opposite corners of the fuel are blackened on the top. This indicates that smoldering passed from the bottom to the top of the fuel bed at these locations. The absence of blackening at other locations on the top surface suggests that the fuel bed was barely able to support smoldering at this heated plate temperature. Even so, glowing combustion was observed during this experiment.

The corresponding photographs for a fuel bed placed on the heated plate at 310 °C are shown in Figure 58. The blackening on the bottom shows that the fuel in contact with the plate was pyrolyzed. Unlike for the higher plate temperature cases shown in Figure 57, the top of fuel bed is only lightly browned in the center showing that smoldering for this fuel bed did not progress to the top from the bottom surface. The light brown area is either due to deposited smoke or limited pyrolysis due to the thermal plume that developed within the fuel bed.

The observations above suggest that if non glowing smoldering developed in these fine Florida grass fuel beds, it would transition to glowing combustion. This was not the case at the lower heated plate temperatures for the other grasses discussed above, for which non glowing combustion was observed without transition to glowing combustion.

The tops and bottoms of fuel beds for which glowing combustion was not observed in the presence of a wind had similar appearances. Figure 59 shows an example for the heated plate

Figure 58. Bottom (left) and top (right) views of a fine Florida grass fuel bed are shown after the fuel was removed from the heated plate held at 310 °C with no wind applied. Glowing combustion was not observed.

Figure 59. Bottom (left) and top (right) views of a fine Florida grass fuel bed are shown after the fuel was removed from the heated plate held at 290 °C with a low wind applied. Glowing combustion was not observed.

held at 290 °C with a low wind. A rectangular area of the bottom fuel surface is blackened, with unblackened bands visible along the upstream (wider band on right side of image) and the downstream edges. As discussed earlier, convective cooling due to wind flow over the fuel and the thermal plume formed by the rising heated gases seemed to cool the fuel and limit reaction at

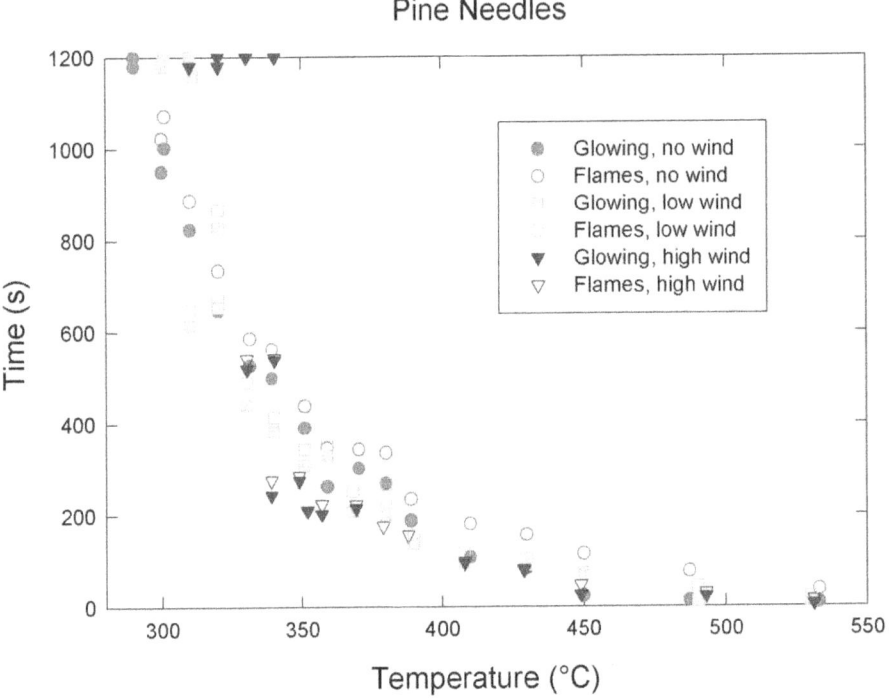

Figure 60. Ignition times for glowing combustion and flaming are shown as a function of temperature for pine needle fuel beds applied to a heated plate. Results are included for no wind, low wind, and high wind cases.

the upstream and downstream edges. The top of the fuel bed appears to be completely unaffected by the heating except for a narrow light brown band running perpendicular to the wind flow direction just downstream of the center. This band is either due to deposited smoke or light pyrolysis by the thermal plume.

4.7. Pine Needles

Both heated plate and cone ignition experiments were done for this fuel.

4.7.1. Heated Surface Ignition Results

The times required for the appearance of glowing combustion and flaming are plotted as a function of heated plate temperature for pine needle fuel beds exposed to no wind, low wind, and high wind in Figure 60. Results for the three wind conditions fall on well defined curves with the data with no wind requiring longer times for a given plate temperature and the measurements with a high wind requiring the shortest periods. The relative separations of the curves are somewhat less than observed for the grasses discussed above, but similar to those for shredded newsprint (see Figure 24).

Figure 61.   Bottom (left) and top (right) views of a pine needle fuel bed are shown after the fuel
was removed from the heated plate held at 290 °C with no wind applied. No
glowing combustion was observed.

There are several differences between the pine needle results and those for the grass samples
discussed earlier. Perhaps the most important is that the transition from glowing combustion to
flaming was observed for all samples that developed glowing combustion, irrespective of wind
condition. All of the grasses were unlikely to flame without an applied wind, and the May tall
fescue resisted flaming even in the presence of a wind. Interestingly, the shredded newsprint
flaming behavior was similar to that for pine needles.

In contrast to the grass findings where the fuel beds developed glowing combustion at lower
plate temperatures when a wind was present, the high wind condition resulted in the highest plate
temperatures for which the pine needles did not begin glowing. Glowing combustion for the
pine needle fuel beds no longer occurred in the presence of a high wind for plate temperatures
between 330 °C and 340 °C. Glowing ignition in the presence of a low wind was not observed
below temperatures of about 310 °C, while plate temperatures had to fall below 300 °C before
glowing combustion did not develop when no wind was present. Shredded newsprint is the only
other fuel that showed a similar dependence on wind, but the range of temperatures over which
the transition from glowing to non glowing occurred for the three wind conditions was somewhat
smaller, taking place from about 350 °C to 330 °C.

Another difference between the shredded newsprint and pine needle fuel beds and the grass fuel
beds is the time required for glowing combustion to develop at the low ends of the heated plate
temperature ranges. For the grasses these times ranged between 200 s and 600 s, with the
Florida fine grass responsible for the longest times. For the shredded newsprint and pine needles
the maximum periods observed approached 1000 s.

Figure 61 shows photographs of the bottom and top of one of the pine needle fuel beds after
removal from the heated plate held at a temperature of 290 °C in the absence of wind. The

Figure 62. Bottom (left) and top (right) views of a pine needle fuel bed are shown after the fuel was removed from the heated plate held at 340 °C with no wind applied. No glowing combustion was observed.

bottom of the fuel bed is blackened, indicating that some pyrolysis of pine needles in contact with the heated plate took place. Smoldering clearly did not spread from the bottom to the top of the fuel bed since there is little, if any, discoloration of the fuel on the top surface. Since fuel beds placed on the heated plate held at temperatures only 10 °C higher developed glowing combustion and flaming, this suggests that if non glowing smoldering did start in these fuel beds, it always transitioned to glowing combustion and ultimately flaming.

Figure 62 shows the bottom and top of a fuel bed that did not develop glowing combustion when placed on the heated plate held at 340 °C with the high wind. A narrow band on the bottom of the fuel bed is blackened on the downstream side. There is little indication of pyrolysis on the upstream side. This is an indication that convective cooling of the fuel has limited reaction at these locations despite an increased air supply. Smoldering clearly did not develop within the fuel bed since the top of the fuel bed shows only a hint of discoloration at the downstream edge. Photographs for other low and high wind cases where glowing combustion did not occur have similar appearances, even though in some the bands of discoloration on the tops of the fuel beds were more pronouced. These observations are consistent with those discussed earlier for other fuel beds that did not smolder in the presence of wind.

4.7.2. Radiative Heating Ignition Results

Plots of time to flaming ignition and maximum HRR as functions of AHF are included in Figure 63 from cone calorimeter measurements with pine needles. The data span a range of AHFs from 15 kW/m^2 to 50 kW/m^2. Flaming ignition was observed over this entire range of AHF. Unfortunately, measurements are not available for lower AHFs.

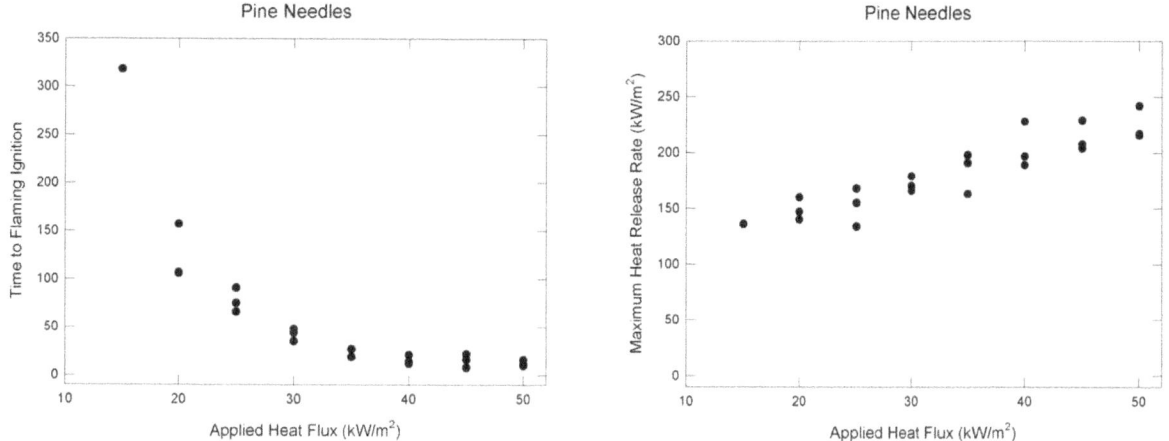

Figure 63. Values of time to flaming ignition and maximum observed heat release rate are plotted as functions of applied heat flux for pine needles.

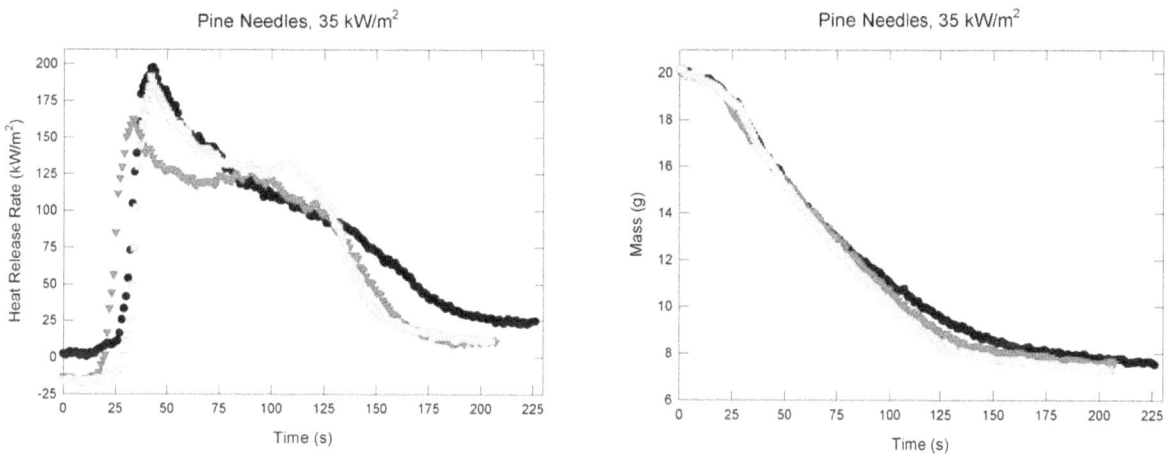

Figure 64. Heat release rates and sample masses are plotted as a function of time for three experiments in which a 35 kW/m² heat flux was applied to pine needles.

Less than 30 s was required for ignition with AHFs equal to or greater than 35 kW/m², and the ignition times varied little with AHF over this range. For AHFs around 30 kW/m² the ignition times began to increase more rapidly with decreasing AHF. The general dependence of ignition times on AHF is similar to those observed for shredded newsprint, May tall fescue, and cheat grass, but for these fuels the rapid increases in ignition times began when the AHF was on the order of 20 kW/m², and the rates of increase with decreasing AHF were higher.

Even though the dependence of the ignition times on AHF is highly non linear, the values of maximum HRR appear to have a nearly linear dependence, falling with decreasing AHF over the entire AHF range. This behavior is expected for fuels where the pyrolysis rate during flaming combustion is proportional to the AHF.

Additional insight into the ignition behavior of pine needles is provided by considering the time behaviors of the HRR and mass loss. Figure 64 shows results for three experiments with an AHF

65

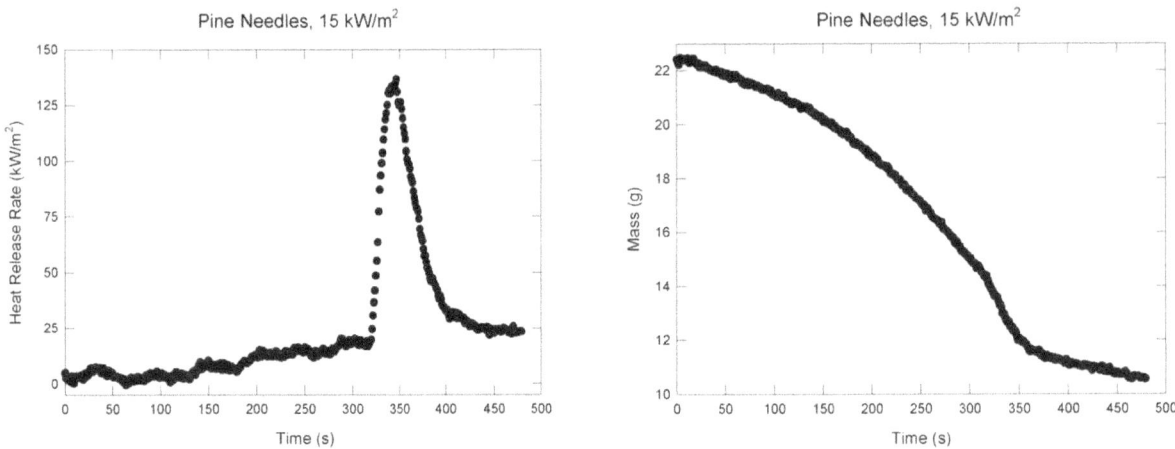

Figure 65. Heat release rates and sample masses are plotted as a function of time for an experiment in which a 15 kW/m^2 heat flux was applied to pine needles.

of 35 kW/m^2. Short induction periods were present at the beginning during which there were relatively low mass loss rates and very little heat release. After the induction period, there were abrupt increases in the rates of mass loss and the HRRs rose very rapidly to their maximum values. Flaming combustion lasted roughly 125 s, before the HRRs and mass loss rates decreased rapidly to levels expected for smoldering and then decreased slowly.

Time plots for an AHF of 30 kW/m^2 have a similar appearance to those in Figure 64, with the exception that the induction periods increased substantially, varying from about 35 s to 50 s, and the HRRs increased slowly to values on the order of 20 kW/m^2 before the rapid increase associated with the transition to flaming. As the AHFs were reduced further the induction periods continued to increase.

Figure 65 shows the HRR and mass loss curves for an experiment with an AHF of 15 kW/m^2. There was a period of roughly 125 s at the start during which the HRR was very close to zero and mass was being loss relatively slowly. After the induction period the HRR began to increase slowly and the slope of the mass loss curve also increased slowly. After about 200 s the HRR had increased to roughly 20 kW/m^2, and there was a sudden transition to flaming.

These data reveal that for low AHFs there are at least three distinct processes involved in the ignition of pine needles. The first is a period during which the fuel mass decreases, but there is little if any heat release. The mass loss is likely due to drying of the fuel and non oxidative fuel pyrolysis. During the second phase the fuel surface begins to oxidize and surface temperatures rise slowly. The increasing surface temperatures lead to more intense pyrolysis since the mass loss rates increase and smoldering spreads. It is unclear if glowing combustion developed in the cone experiments, but observations from the heated plate experiments suggest that non glowing smoldering is not likely to play an important role for this fuel. For these fuel beds, transition to flaming seems to take place when the measured HRRs reach a value of approximately 20 kW/m^2.

66

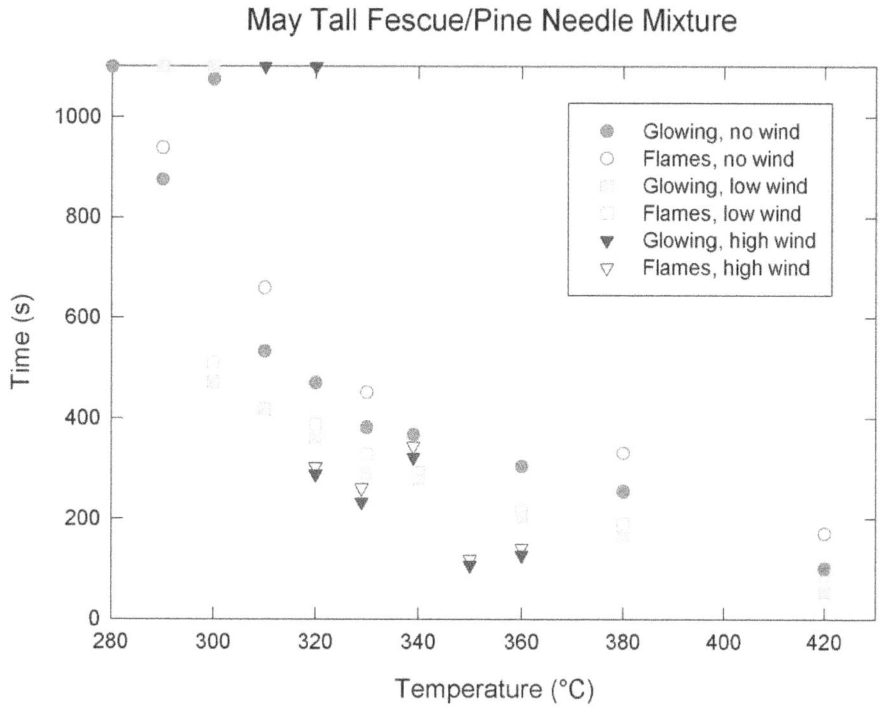

May Tall Fescue/Pine Needle Mixture

Figure 66. Ignition times for glowing combustion and flaming are shown as a function of temperature for May tall fescue/pine needle mixture fuel beds applied to a heated plate. Results are included for no wind, low wind, and high wind cases.

4.8. May Tall Fescue/Pine Needle Mixture

Only heated plate ignition experiments were performed for the May tall fescue/pine needle mixture. Figure 66 shows the times required for the development of glowing combustion and flaming plotted as a function of plate temperature for no, low, and high wind cases. The results for the different wind conditions fall on well defined curves with ignition times that increase as the plate temperature decreases. Generally, for a given plate temperature, the longest ignition times were observed for no wind cases and the shortest are for high wind cases.

The two fuels in the mixture were chosen because they represented two extremes in the ignition behaviors observed for the natural fuels considered in this study. May tall fescue (see Figure 33) had a low tendency to flame when placed on a heated plate for any of the wind conditions tested, while the pine needles (see Figure 60) transitioned to flaming whenever glowing combustion developed. The application of a wind resulted in larger reductions in glowing combustion ignition times with a given plate temperature for the May tall fescue. In the presence of a wind the development of glowing combustion in the May tall fescue was observed for lower plate temperatures than when a wind was not present. The opposite was true for pine needles. The times required for the appearance of glowing combustion at a given plate temperature were typically much longer for the pine needles as compared to the May tall fescue.

67

Comparison of Figure 66 with Figure 33 and Figure 60 suggests that a 50 %/50 % mixture by mass of May tall fescue and pine needles resulted in fuel beds with ignition behaviors that were intermediate between those for the individual fuels, but which also had some properties that might be considered less desirable than those for the individual fuels alone.

The results in Figure 66 show that the fuel mixture was much more likely to flame than the May tall fescue. Transition from glowing combustion to flaming was observed in all cases with an applied wind and in just over half of the cases without wind. It is noteworthy that the tendency for the mixture to flame is greater than for all of the grasses tested. For the experiments without an applied wind, there is no clear correlation of the appearance of flames with plate temperature, suggesting that the probability of transition to flaming for this fuel mixture is related to the flammability of the gases generated by pyrolysis and not to the specific local ignition conditions.

For the lower plate temperatures where both of the indivual fuels developed glowing combustion, the ignition times for the mixture tended to lie between those observed for the individual fuels, but to fall somewhat closer to those for the May tall fescue. As an example, consider the low-wind experiments with plate temperatures around 310 °C. Glowing combustion required over 600 s to develop for the pine needles, while the comparable time for the May tall fescue was between 250 s and 275 s. The ignition time for the mixture was just over 400 s. This behavior suggests that the May tall fescue first develops non glowing smoldering with the heat generated by the surface oxidation being transferred to both non smoldering May tall fescue and pine needles. Eventually surface temperatures on a portion of the mixed fuel reach a point where transition to glowing combustion occurs. This glowing combustion spreads and increases the heat release until it is sufficient to ignite the pyrolysis gases generated by the smoldering, and transition to flaming takes place.

For the pine needles the lowest plate temperature where glowing and flaming ignition was observed was 300 °C without a wind present, and roughly 1000 s was required for glowing combustion to develop. For the same plate temperature the May tall fescue/pine needle mixture with a low wind required just under 500 s to begin glowing and rapidly transition to flaming. The lowest plate temperature that produced a glowing and/or flaming ignition for all of the fuels studied during this study was the 290 °C ignition of the fuel mixture in the absence of the wind. As discussed earlier, non glowing smoldering was observed for the May tall fescue for the lowest plate temperature, 302 °C, run without an applied wind. It is likely that non glowing smoldering of the May tall fescue within the fuel mixture generated the initial heat necessary to ignite glowing combustion in the fuel mixture.

The conclusion that non glowing combustion of the May tall fescue component is responsible for the initial heat release in the fuel mixture is supported by the photographs of the bottom and top of the fuel bed shown in Figure 67 for an experiment with the heated plate held at 280 °C and no applied wind. No glowing combustion was observed in this experiment. The bottom of the fuel bed is blackened, indicating significant fuel pyrolysis had occurred. On the top there is a circular dark brown area indicating that some pyrolysis of the fuel mixture took place here as well. Given the low plate temperature, pyrolysis on the top of the bed would not have been expected unless a non glowing smolder front had passed upward through the fuel bed. It has been shown that smoldering in the pine needle beds studies here always involved glowing combustion.

Figure 67. Bottom (left) and top (right) views of a May tall fescue/pine needle mixed fuel bed are shown after the fuel was removed from the heated plate held at 280 °C with no wind applied. No glowing combustion was observed.

At lower plate temperatures the effect of applying a wind to the fuel mixture was to decrease the likelihood of ignition since glowing combustion was observed for lower plate temperatures in the absence of a wind than with wind present. This dependence on wind is the same as observed for pine needles and suggests this component of the fuel mixture dominated this particular behavior.

The importance of convective cooling of the mixed fuel by an applied wind on the ignition behavior is demonstrated by the images shown in Figure 68 for a fuel bed removed from the heated plate held at 320 °C in the presence of a high wind. On the bottom of the fuel bed there is a small blackened area on the downstream half of the bed. Fuel pyrolysis has clearly occurred here. There is some discoloration of the fuel bed along a narrow band lying perpendicular to the wind direction near the downstream edge suggested smoke deposition from or some light pyrolysis by the thermal plume passing through the fuel bed. Similar observations for other fuels have been attributed to convective cooling of the fuel near the heated plate.

4.9. Preliminary Tests with Leaves

A limited number of ignition experiments were carried out on the heated plate using three types of dried leaves. All of these tests were run with heated plate temperatures of 381 °C. This temperature was chosen because the experiments discussed above indicated that this temperature is high enough to ensure that glowing combustion will develop in a reasonably short period of time.

A single experiment was done for 8 g of boxwood leaves placed on the heated plate without a wind. Glowing combustion was observed 90 s after the leaves were applied, but transition to

Figure 68. Bottom (left) and top (right) views of a May tall fescue/pine needle mixed fuel bed are shown after the fuel was removed from the heated plate held at 320 °C with a high wind applied. No glowing combustion was observed.

flaming did not occur. Photographs of the fuel bed showed blackening on the bottom, a circular blackened area on the top, and an area of unreacted fuel around the outside edge of the bed.

Two experiments were run with 5.5 g of elm leaves. When a low wind was used, glowing combustion and flaming appeared nearly simultaneously after 93 s. Smoke was observed coming from the fuel bed along a line perpendicular to the wind flow direction located about 2/3 of the way across the bed from the upstream edge. Photographs show that most of the fuel was blackened. For the second experiment no wind was applied. Glowing combustion developed 253 s after fuel application, but no transition to flaming was observed. The glowing combustion spread quickly to cover most of the fuel bed.

Two repeats of an experiment with the oak leaves in the presence of a low wind were done. For the first test 6.0 g of leaves were used. Neither glowing combustion nor flaming were observed, but heavy smoke was observed along a line perpendicular to the wind flow direction, indicating that pyrolysis was occurring. Photographs after the experiment showed that the bottom was heavily blackened, but the top appeared to be unaffected. This suggests that sustained smoldering did not occur. For the second experiment the mass of oak leaves was increased to 9.4 g. Glowing combustion was observed after 216 s, and flames appeared 11 s later. Both ignitions occurred near the surface of the heated plate. The fuel bed was completely blackened, and there was some ash formation.

The experiments with leaves are too few in number to draw many definite conclusions, but it is clear that leaves can ignite and flame for plate temperatures around 380 °C. The results also

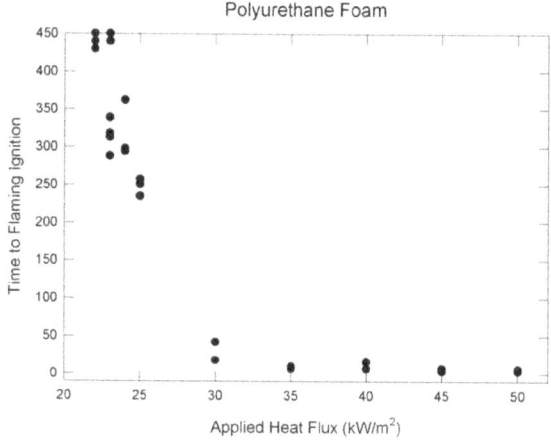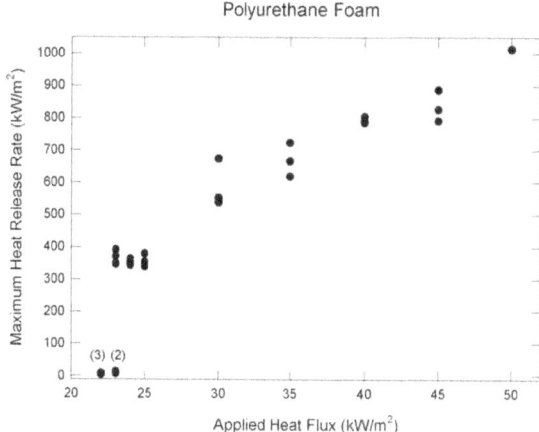

Figure 69. Values of time to flaming ignition and maximum observed heat release rate are plotted as functions of applied heat flux for polyurethane foam. Numbers in parentheses refer to the number of repeated measurements.

suggest that the application of a wind will have similar effects in enhancing ignition to those for the fuels discussed earlier.

4.10. Radiative Ignition of Non Fire-Retarded Flexible Polyurethane Foam

Non fire-retarded flexible polyurethane was included in the study as an example of a plastic material that might be found in an area where outdoor power equipment is stored. Experiments were done only in the cone calorimeter.

Figure 69 shows measured times to flaming ignition and maximum HRRs as a function of AHF. For AHFs between 30 kW/m^2 and 50 kW/m^2 the times required for flaming ignition increased slowly as the AHF decreased. When the AHF was decreased to 25 kW/m^2 the ignition times suddenly jumped to values on the order of 250 s. When the AHF was reduced to 24 kW/m^2 the ignition times increased by about another 50 s. Flaming was observed in only 4 of 6 tests with a further reduction of 1 kW/m^2 in AHF. Maximum ignition times for the cases that flamed were roughly 325 s. When the AHF was set to 22 kW/m^2 no flaming ignitions were observed for three tests.

Despite the highly non linear dependence of the ignition times on AHF, there was a nearly linear decrease in the maximum HRRs, with values dropping from about 1000 kW/m^2 for an AHF of 50 kW/m^2 to around 350 kW/m^2 for the lowest AHFs where ignition occurred.

Time resolved measurements of HRR and fuel mass reveal that the ignition and burning behaviors of the polyurethane foam were different from those observed for the other fuels. Figure 70 shows plots for an AHF of 50 kW/m^2. Following a brief induction period during which the HRR grew slowly and the mass loss rate was low, the HRR and mass loss rate grew

71

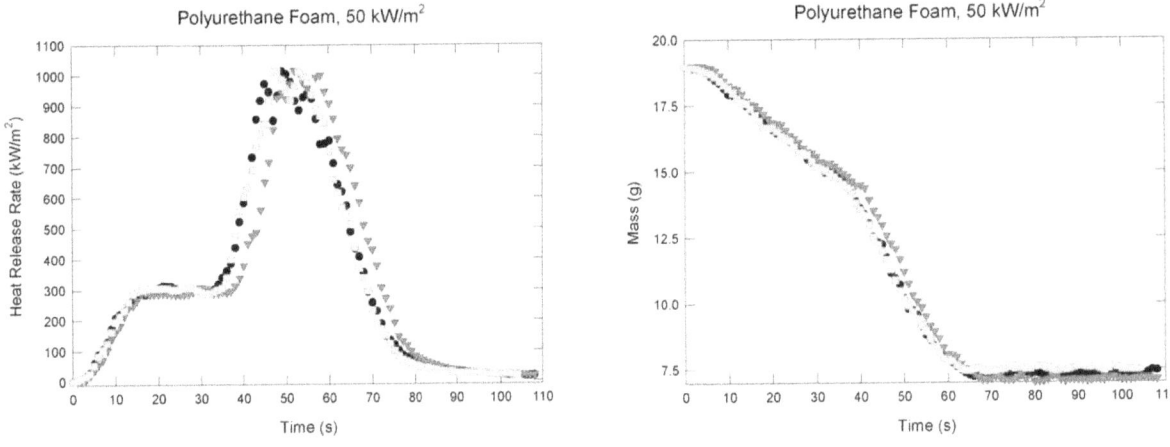

Figure 70. Heat release rates and sample masses are plotted as a function of time for three experiments in which a 50 kW/m² heat flux was applied to polyurethane foam.

rapidly to much higher plateau values that lasted about 25 s. About 40 s after exposing the sample to the AHF the HRR rate began to increase again, rising from about 300 kW/m² to a peak of around 1000 kW/m². The mass loss rate during this second burning phase was much higher than during the first. Following the second burning period the HRR fell rapidly to values close to zero, and the mass did not change appreciably.

Visual observations of flames spreading across a similar polyurethane foam provide a plausible explanation for the observed HRR behavior. As the flames spread onto unburned foam, the foam first darkened and began to shrink. As the foam continued to burn and shrink it generated a brown liquid that began to collect. Once the foam had fully collapsed, the liquid remained as a pool, which then burned as a pool fire. The first heat release peak in Figure 70 is likely associated with burning of the expanded foam. As the foam burned it was forming a liquid pool in the bottom of the sample pan, which eventually ignited and resulted in the second HRR peak.

The HRR and fuel mass time behaviors for tests with AHFs between 30 kW/m² and 50 kW/m² had similar appearances to those in Figure 70. As the AHF was reduced the induction period tended to increase. For an AHF of 30 kW/m² the induction times began to vary significantly, ranging from 15 s to 35 s.

Figure 71 includes plots showing HRR and fuel mass as a function of time for three experiments with an AHF of 25 kW/m². These curves have a very different appearance from those with an AHF of 50 kW/m² shown in Figure 70. There is a very long induction period during which a large fraction of the fuel mass is loss without significant heat release. This suggests that non oxidative pyrolysis of the foam was taking place. After periods ranging from 230 s to 250 s, the HRR began to rise quickly and flaming was present for a short period of time. During flaming the mass loss rates increased substantially, having slopes comparable to those present during the second burning phase with an AHF of 50 kW/m². Taken together, these observations suggest that the pyrolyzate generated directly from the foam was not ignited when a AHF of 25 kW/m² was used, but that the liquid pool that was generated by this pyrolysis did eventually ignite.

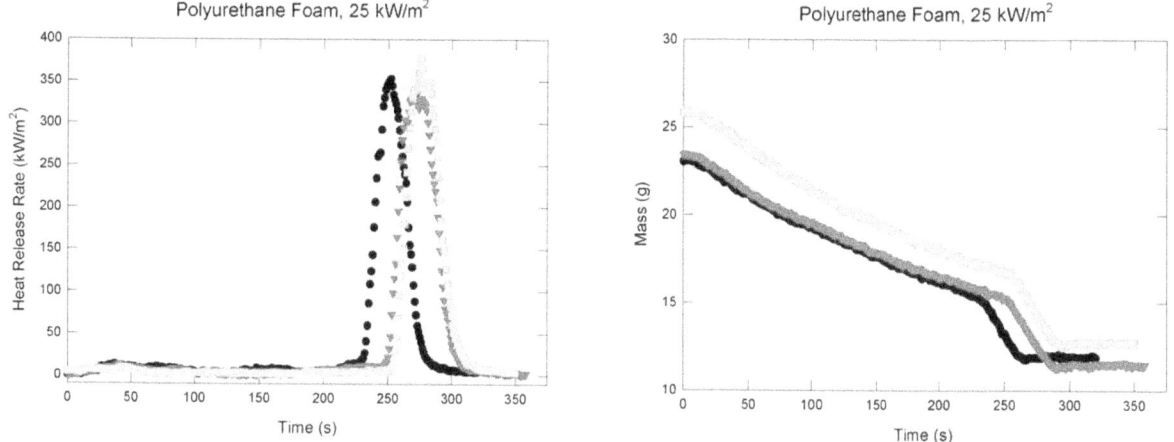

Figure 71. Heat release rates and sample masses are plotted as a function of time for three experiments in which a 25 kW/m^2 heat flux was applied to polyurethane foam.

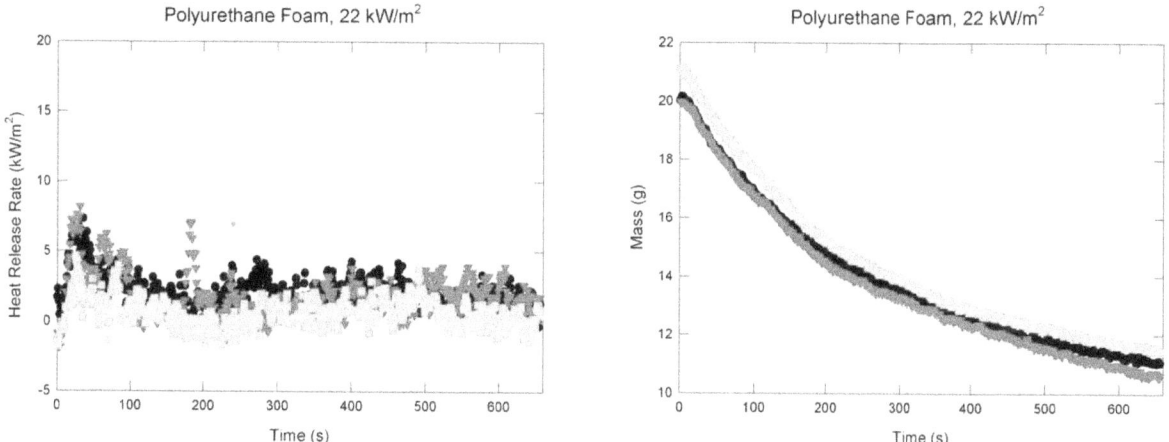

Figure 72. Heat release rates and sample masses are plotted as a function of time for three experiments in which a 22 kW/m^2 heat flux was applied to polyurethane foam.

During these experiments there was concern that the ignition of the foam might be due to the presence of the cone heater acting as a pilot for the pyrolysis gases being generated. This possibility was checked by making a video recording of the ignition. It was clear from analysis of this video that ignition occurred within the sample holder and then spread rapidly upwards.

Plots (not shown) of HRRs and fuel mass as functions of time for the three experiments with an AHF of 24 kW/m^2 and the four experiments that ignited with an AHF of 23 kW/m^2 are similar to those shown in Figure 71 with induction times that determine the times to ignition shown in Figure 69.

Figure 72 shows similar plots for an AHF of 22 kW/m^2 for which ignitions were not observed. The HRR plots suggest that there were brief periods during which some oxidation was occurring, but overall the HRRs were close to zero over the entire periods. On the other hand, a large fraction of the fuel mass was loss. This indicates that non oxidative pyrolysis was taking place in cases where no ignition was observed. At the end of these experiments the inside of the sample holder was covered with a thin layer of solid black material. This was likely the remnants of the liquid material that formed during the pyrolysis of the polyurethane foam.

Due to the complex reaction behavior of the foam that led to the formation of a liquid and that left a hard-to-remove residue, the decision was made to not risk testing this material on the heated plate.

5. Summary and Discussion

The purpose of this investigation was to provide experimental findings that can be used to better estimate the potential for ignition associated with the heated exhaust surfaces present during the operation of OPE. Due to the application, emphasis was placed on natural fuels often found in the outdoor environment. Scenarios involving direct contact of fuel with a heated surface and radiative heating of fuel placed near a heated surface have been considered. A number of important parameters have been identified and varied during the investigation, including fuel type, heated surface temperature and intensity of applied radiative heat flux, and the absence or present of an imposed wind.

An important characteristic of most of the fuels considered is the ability to smolder. The primary source of information on smoldering behavior of the fuels used in this investigation is derived from the heated plate experiments. While for a limited number of experiments non glowing smoldering without a transition to glowing was observed (based on blackening of the fuel), it was far more common for glowing combustion to develop. As discussed earlier, the development of glowing appears to be a requirement for the ignition of flames in these fuel beds. The times required for the appearances of glowing and flaming are the primary data for the experiments reported here.

As for all experiments, there are uncertainties associated with the measurements that need to be assessed. As discussed in Section 4.1, some uncertainty in ignition times was introduced during visual observation of glowing combustion. In some cases the initial glowing was hidden within the fuel bed and glowing was not observed until a later time. It is believed in these cases that the time periods between initial glowing and the detection of combustion (either glowing or flaming) were relatively short and have little or no effect on the general conclusions drawn from the experiments.

Due to the stochastic nature of the fuel beds and the natural variation in the tendency of individual fuel elements to ignite, uncertainties in measured ignition times for a given heated plate temperature or radiative heat flux are inevitable. The only effective way to quantitatively characterize these variations would be to perform sufficient experiments to allow statistical analysis. It was not possible to perform the large number of experiments required within the scope of the current effort. In the absence of such data, a qualitative approach was adopted in which the degree of collapse of experimental data onto well defined curves is used as a qualitative basis for assessing uncertainty.

74

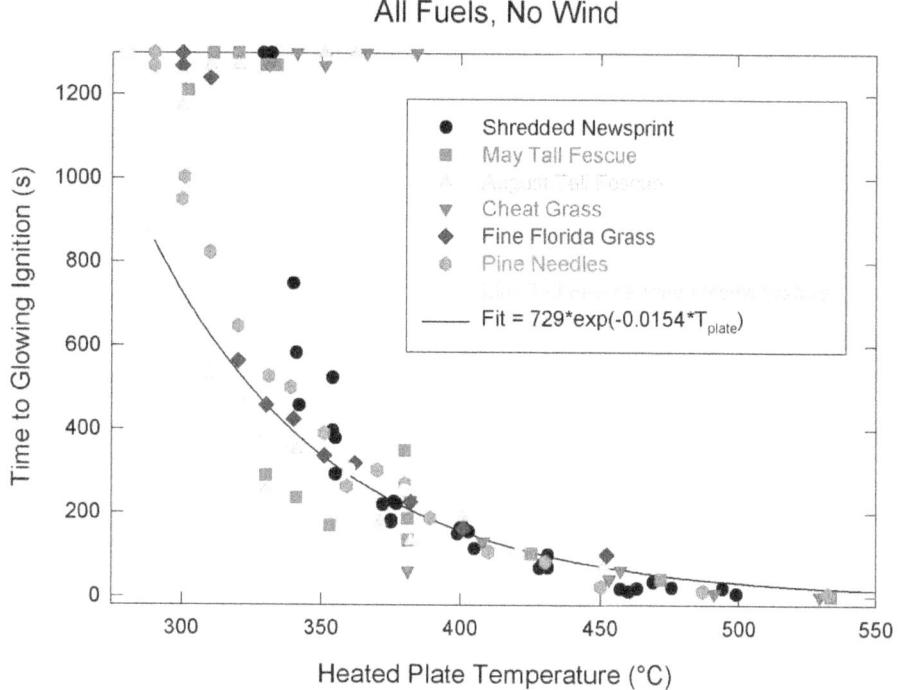

Figure 73. Data for time to glowing ignition with no applied wind are plotted as a function of heated plate temperature for seven fuels. Data near the upper temperature axis represent experiments for which ignition did not take place. The solid curve is the result of a least squares fit assuming an exponential dependence on heated plate temperature for all data where ignition was observed.

Similar considerations apply to determinations of the minimum heated plate temperatures or radiative heat fluxes required to ignite a given type of fuel bed. A detailed study would be expected to reveal an ignition probability curve in which the probability of ignition would increase from near zero to near one as temperature was increased over some range. Mapping out such a probability curve would have required far too many experiments. For this work an approach was adopted in which heated plate temperatures or radiative heat fluxes were reduced in discrete steps. Usually, but not in all cases, the step sizes were such that well defined demarcations between experiments with and without ignition were identified. Lower temperature ignition limits can be identified from these experiments, but due to the stochastic nature of the processes it is prudent to assign an uncertainty of at least one step size to the results.

Results for individual fuels have been discussed in Section 4. In order to assess the roles of fuel type, heated plate temperature, and wind for the surface ignition experiments, ignition time data (the shorter of glowing or flaming) for all of the fuels tested have been plotted together on a single plot. Figure 73 shows the result for experiments without an applied wind. For comparison purposes, a least squares curve fit assuming the fall off of ignition times with increasing plate temperature has an exponential dependence was fit to all of the data for which ignition was observed. The result of the fit,

$$t_{glowing,fit,none} = 729 \times e^{-0.0154 T_{heated\ plate}},$$ (2)

75

where $t_{glowing,fit,none}$ is the calculated time to ignition and $T_{heated\ plate}$ is the heated plate temperature is shown as a solid curve in Figure 73 . The experimental data fall on a wide band on either side of the curve fit.

Data for each fuel shows a similar dependence on plate temperature, with the ignition time increasing with decreasing plate temperature and the rate of increase accelerating. Comparison with the exponential curve fit included in Figure 73 shows that an exponential dependence on plate temperature provides a good approximation for the temperature dependence.

Closer inspection of Figure 73 shows that data for individual fuels lie on well defined curves that show much smaller variations than the consolidated data. These relatively small variations show that variations in measurements for individual fuels are much smaller than those due to variations between fuels. Interestingly, data for all of the fuels seem to lie close together for plate temperatures above 400 °C, with larger relative differences between fuels appearing for lower heated plate temperatures.

Results for the May and August tall fescue grasses fall close together and lie well below the curve fit, data for fine Florida grass and the May tall fescue/pine needle mixture fall close to the curve fit, and the shredded newsprint and pine needle data appear well above the curve. It is difficult to assess the behavior of cheat grass because it did not ignite for temperatures below 380 °C. Recall that transition to flaming in the absence of a wind was unlikely for May tall fescue, August tall fescue, and cheat grass, while the May tall fescue/pine needle mixture transitioned to flaming in over half of the tests, and the pine needles and shredded newsprint were likely to flame for all wind conditions. The probability of transition to flaming in the absence of wind seems partially correlated with the time-to-ignition variations observed between fuels, with a shorter time to ignition correlating with a lowered probability of transition to flaming.

The data at the top of the plot in Figure 73 represents experiments for which glowing or flaming ignition was not observed. Minimum ignition temperatures show distinct fuel-type dependencies. The highest minimum ignition temperature (380 °C) was for cheat grass. This was followed by the August tall fescue (371 °C) and May tall fescue (340 °C). Recall that even though glowing combustion was not observed just below the transition temperature for these two grasses, non glowing smoldering was observed down to temperatures around 300 °C. The minimum heated plate ignition temperature (340 °C) for shredded newsprint was similar to that for May tall fescue. Next was the fine Florida grass with a minimum ignition temperature of around 320 °C. The minimum ignition temperatures were 300 °C for pine needles and 290 °C for the May tall fescue/ pine needle mixture. In general, the four grasses seem to have the highest minimum ignition temperatures, but there is no clear correlation between the minimum ignition temperatures and the relative ordering of the ignition time curves.

Results of the heated plate ignition experiments with a low wind for the seven fuels are plotted in Figure 74. The data fall within a band and have a similar temperature dependence to those for the no wind case shown in Figure 73. A least squares curve fit to an exponential gave

$$t_{glowing,fit,low} = 106,000 \times e^{-0.0176 T_{heated\ plate}}, \tag{3}$$

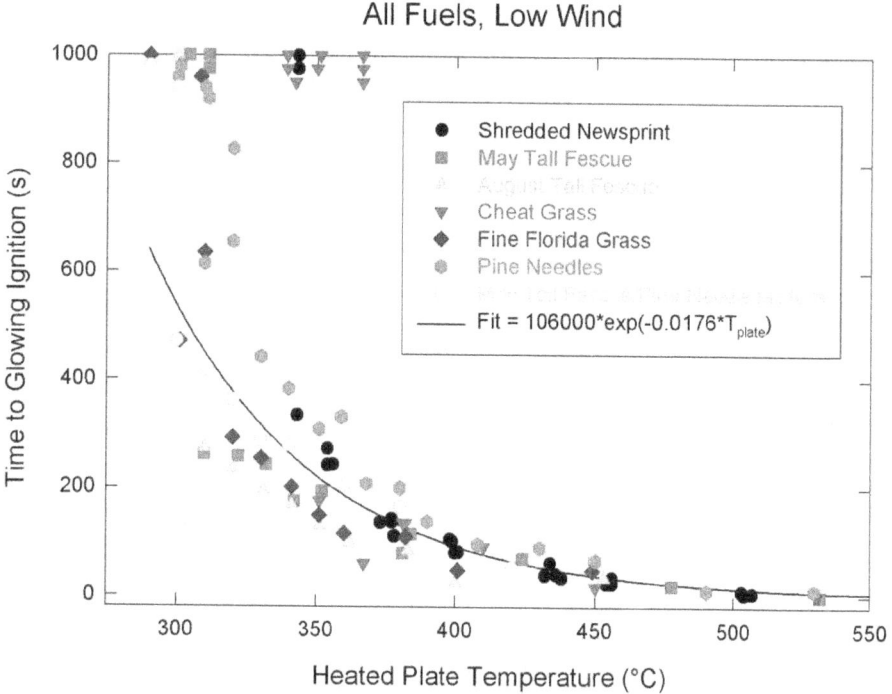

Figure 74. Data for time to glowing ignition with a low applied wind are plotted as a function of heated plate temperature for seven fuels. Data near the upper temperature axis represent experiments for which ignition did not take place. The solid curve is the result of a least squares fit assuming an exponential dependence on heated plate temperature for all data where ignition was observed.

and is shown as a solid curve in Figure 74. Similar to the no wind case, the data for individual fuels fall on bands with fluctuations that are smaller than the differences between the fuels.

As for the no wind case, the data lie fairly close together for heated plate temperatures greater than 400 °C and begin to diverge at lower temperatures. The data for May tall fescue, August tall fescue, and fine Florida grass are close together and lie below the curve fit. The results for the May tall fescue/pine needle mixture lie close to the calculated curve, while those for shredded newsprint and pine needles fall above. As for the no wind case, it is difficult to assess cheat grass because it did not ignite below 380 °C. The locations of the curves for the different fuels relative to the curve fit are similar to those for the no wind experiments, with the exception that the data for fine Florida grass lies somewhat closer to those for May tall fescue and August tall fescue.

Minimum heated plate ignition temperatures for cheat grass (380 °C), shredded newsprint (340 °C) pine needles (310 °C), and the May tall fescue/pine needle mixture (300 °C) were similar to those observed without an applied wind. May tall fescue, August tall fescue, and fine Florida grass each ignited for plate temperatures as low as 310 °C. For all three grasses this value is lower than observed without a wind present.

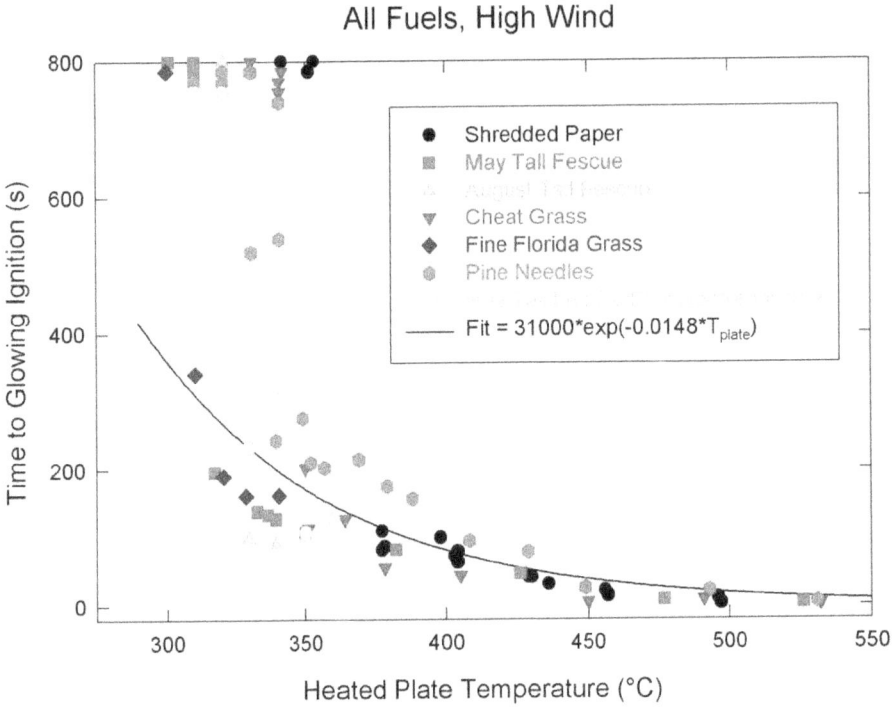

Figure 75. Data for time to glowing ignition with a high applied wind are plotted as a function of heated plate temperature for seven fuels. Data near the upper temperature axis represent experiments for which ignition did not take place. The solid curve is the result of a least squares fit assuming an exponential dependence on heated plate temperature for all data where ignition was observed.

A similar plot of ignition times versus heated plate temperature is shown in Figure 75 for experiments with a high wind. For some of the fuels fewer measurements were done than for the no and low wind cases, but the results follow similar trends. The variations with plate temperature are similar, and a least squares curve fit to an exponential for all of the data with ignition yields

$$t_{glowing, fit, high} = 310,000 \times e^{-0.0148 T_{heated\ plate}},$$ (4)

which is shown in the figure as the solid curve.

As for the no wind and low wind cases, the data for individual fuels fall on well defined curves having smaller fluctuations than the variations between different fuels. The results for May tall fescue, August tall fescue, and fine Florida grass fall close together and lie below the curve. Unlike earlier experiments where the cheat grass did not ignite with a plate temperature below 380 °C, the lowest ignition temperature with a high wind was 350 °C. The data for the cheat grass fall close to the curve fit for all of the data, as do the measurements for the May tall fescue/pine needle mixture. As observed for the other two wind cases, longer ignition times were required for pine needles than indicated by the curve fit. It is not possible to assess the behavior of the shredded newsprint due to the gap of 30 °C between the lowest plate temperature where ignition was observed at 380 °C and the plate temperature where ignition did not occur.

78

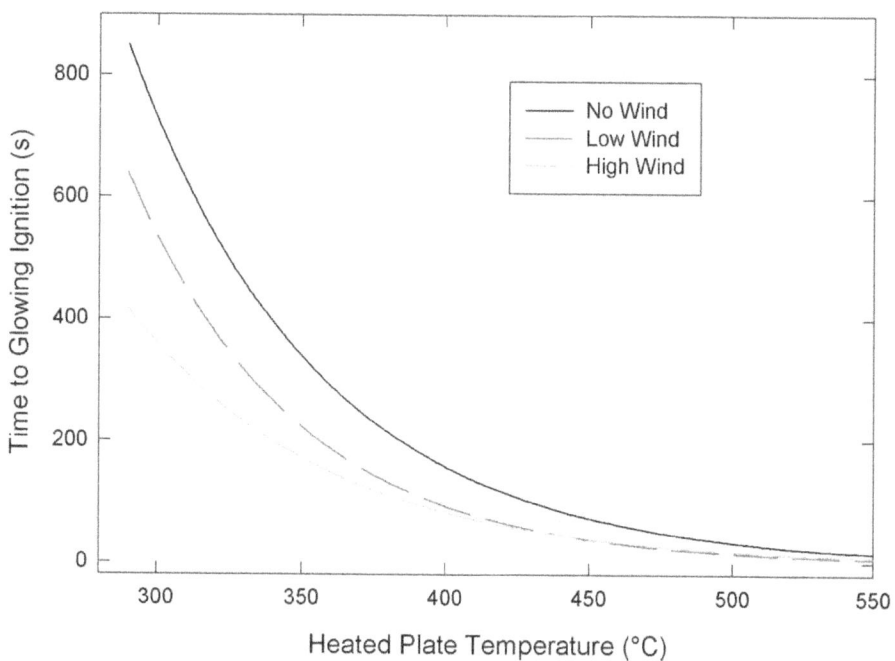

Figure 76. Results of exponential least squares curve fits of experimental ignition times as a function of heated plate temperature for seven fuels are shown for no wind, low wind, and high wind cases.

The highest value of minimum heated plate ignition temperature for these fuels was found for the shredded newsprint, lying between 380 °C and 360 °C. The next was the cheat grass with a minimum ignition temperature of 350 °C, which is about 30 °C lower than observed for this fuel with the other two wind conditions. The minimum ignition temperature for pine needles was 340 °C, but note that this material failed to ignite for some experiments at both 340 °C and 350 °C. The lowest minimum ignition temperatures for the three remaining grasses and the May tall fescue/pine needle mixture fell between 310 °C and 330 °C, with the fine Florida grass having the lowest value.

The data plotted in Figure 73 to Figure 75 show how ignition behaviors depend on the heated plate temperature and fuel for the porous beds studied here. Additional insights into the effects of wind arc obtained by plotting the curve fits given by Equation (2) to Equation (4) on a common plot, as shown in Figure 76. It should be kept in mind that these curve fits are for seven fuels and that there were systematic variations in the data with fuel. Nonetheless, given the similarities in the spreads of data and the similar effects of fuel for the three wind cases, it seems appropriate to use these fits to characterize the effects of wind on surface ignition.

The feature that stands out in Figure 76 is the substantial reduction in ignition times with increasing wind speed for a given heated plate temperature. This effect is particularly pronounced over the lower portions of the temperature range. The dependence on wind is monotonic for the three conditions tested, suggesting that further reductions in ignition times would be likely at higher wind speeds. It should be noted that the two wind velocities used,

1.1 m/s and 2.5 m/s, correspond to 2.5 mph and 5.6 mph. These velocities are comparable to speeds at which OPE operate, and could be induced by equipment motion in still air. Outdoor winds are frequently considerably higher than these levels, so substantial additional reductions in ignition times might be possible for conditions typically encountered in outdoor environments.

The reductions in ignition time with increasing wind as well as qualitative observations concerning the ignition behavior such as initial ignition location and glowing intensity indicate that the primary effect of wind is to increase the amount of oxygen reaching the fuel surface. Since surface oxidation rates for temperatures high enough to generate meaningful reaction rates are generally limited by the rate at which oxygen reaches the surface, the effect is to accelerate the surface reaction rate and decrease the ignition times for both glowing and flaming ignition.

A wind effect that is not revealed by the results plotted in Figure 76 is the potential to convectively cool the fuel and thus slow or prevent ignition. Observed changes in the minimum plate temperatures required for ignition with wind speed and the distributions of blackening on the bottom of fuel beds that did not ignite in the presence of a wind provide evidence that convective cooling affected the surface reaction behavior of the porous fuel beds studied here. Presumably, the observed ignition behavior is determined by a balance of the competing effects of increased oxygen availability and convective cooling. The need to consider these competing effects should be kept in mind when extending the results of this study to wind conditions that were not investigated.

The non piloted thermal radiation ignition behaviors for four of the seven fuels investigated in detail using the heated plate were also studied with the cone calorimeter. Figure 77 shows plots of glowing or flaming ignition time versus AHF for shredded newsprint, May tall fescue, cheat grass, and pine needles. The data plotted near the upper axis represent experiments for which ignition was not observed.

The data for the different fuels fall on well defined curves which have a similar dependence on AHF to the ignition time variations observed with heated plated temperature (see Figure 73 to Figure 76). For the highest AHFs ignition occurs within a few seconds. As the AHF is reduced from the highest values there is a range where the ignition times increase slowly with decreasing AHF. With further reductions in AHF, a point is reached where the measured ignition times begin to rise rapidly. For pine needles the rapid rise begins around 30 kW/m^2 and for the remaining three fuels around 20 kW/m^2. The rate of increase with decreasing AHF is much higher for the three remaining fuels than for pine needles.

The data for the shredded newsprint, May tall fescue, and cheat grass fall close together. These three sets of data were combined, and a least squares fit to an exponential was obtained. The result,

$$t_{ignition, fit} = 5540 \times e^{-0.260 AHF},$$ (5)

is shown in Figure 77 as a solid line. The measurements for the three fuels fall close to the calculated curve.

80

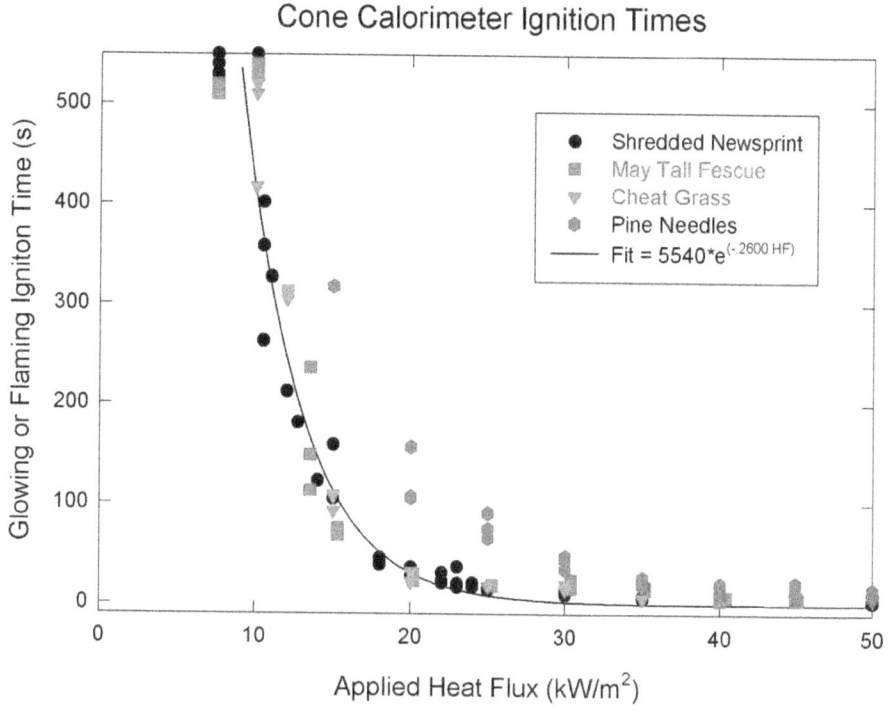

Figure 77. Ignition times due to non piloted radiative heating in the cone calorimeter are plotted as a function of applied heat flux for the four fuels indicated. The solid line shows the results of a least squares exponential curve fit to the combined data for shredded newsprint, May tall fescue, and cheat grass.

Due to the similarities in the response curves, it is worthwhile to consider whether the radiative ignition results can be related to those for the heated plate in some way. One approach is to compare effective temperatures for the radiant heat flux, determined by assuming the thermal radiation is generated by a black body, with corresponding heated plate temperatures. The radiative heat flux (RHF) from an ideal black body is given by

$$RHF = 5.67 \times 10^{-13} * T_{BB}^4, \tag{6}$$

where the *RHF* has units of kW/m² and T_{BB} is the black body temperature in K. Values of temperature derived from Equation (6) are plotted as a function of RHF in Figure 78.

The results in Figure 77 indicate that ignition did not occur for AHFs of 7.5 kW/m² and that the minimum AHF required to ignite smoldering was around 10 kW/m² for shredded newsprint, May tall fescue, and cheat grass. The corresponding black body temperatures for these AHFs are 330 °C and 375 °C. The minimum observed heated plate ignition temperatures for these fuels with no applied wind were 340 °C, 340 °C, and 380 °C. The close agreement between the minimum heated plate ignition temperatures and the black body temperature range over which radiative ignition no longer occurred suggests that the processes determining the minimum ignition conditions are similar for the two types of heating and that the fuel surfaces are being heated to comparable temperatures.

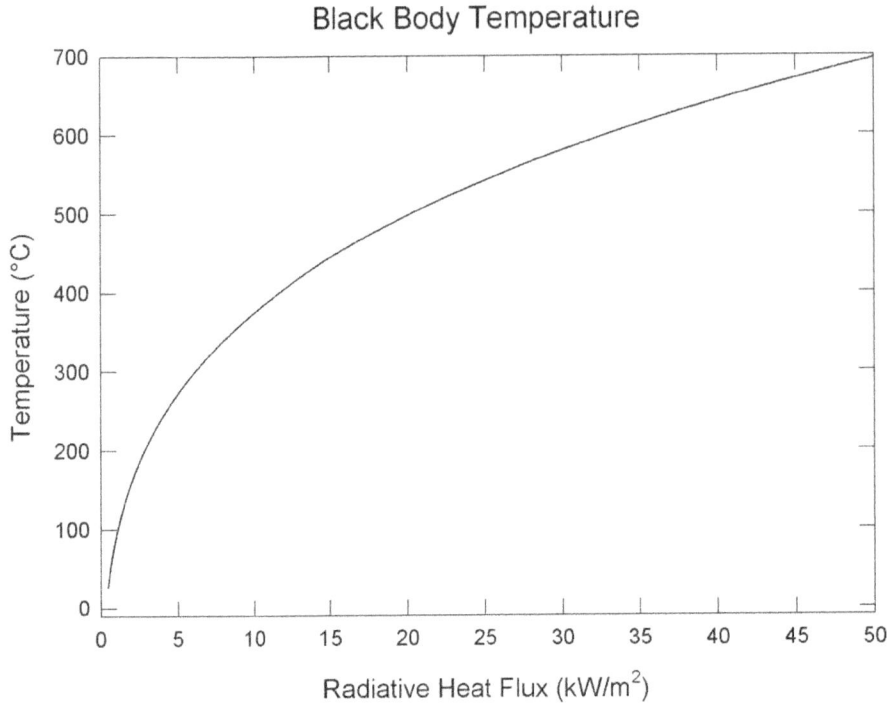

Figure 78. Temperatures are plotted as a function of radiative heat flux assuming that the flux is generated by an ideal black body.

The results in Figure 73 show that ignition times increased slowly with decreasing plate temperature between 550 °C and 450 °C and then began to increase more rapidly with an increasing slope as heated plate temperatures were lowered further. A similar change in slope occurs around 20 kW/m^2 for the time to ignition data in Figure 77. The black body temperature for 20 kW/m^2 is 498 °C. This temperature is somewhat higher than the 450 °C estimate from the heated plate experiments, but is close enough to support the conclusion that the response of these fuels to the two heating types is similar. This is particularly so when it is noted that differences in response should be expected based on configurational differences between the two experiments. The heated plate applied heat at the bottom of the fuel bed, and a buoyant plume, which entrained air through the sides of the fuel bed, was generated. This plume both heats fuel above the plate and provides additional oxygen for surface oxidation. On the other hand, in the cone calorimeter experiment the radiative heating was applied to the top of a fuel bed that was closed on the sides. In this case buoyancy-induced flow by heating near the top surface of the bed is expected to remove heat from the fuel bed, and the absence of air flow into the fuel bed from the sides results in a reduced surface oxidation rate compared to the heated plate experiments. Both of these effects are expected to lengthen ignition times.

The above discussion refers to the ignition behaviors of shredded newsprint, May tall fescue, and cheat grass. The agreement between experiments with the two types of heating is not as good when pine needles are considered. The radiative ignition times for this fuel began to increase around 30 kW/m^2. This corresponds to a black body temperature of 580 °C. Corresponding increases in ignition times for heated plate ignition did not appear until the plate temperature was reduced to around 450 °C. It has not been possible to provide an explanation for this difference.

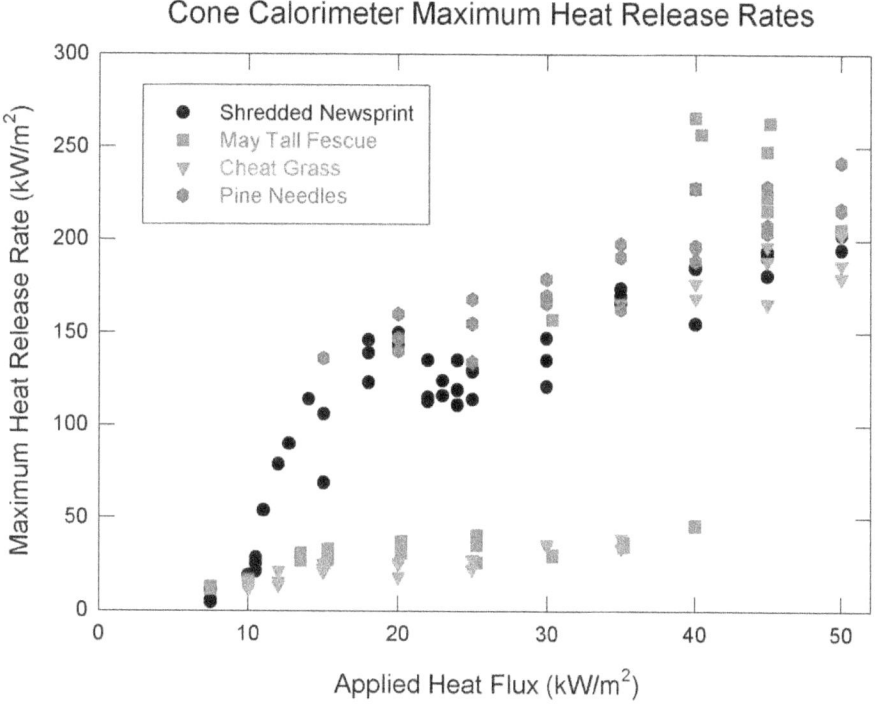

Figure 79. Values of maximum heat release rate are plotted as a function of applied heat flux for the four fuels indicated.

The behavior of maximum HRRs determined in the cone calorimeter experiments provide additional information concerning ignition behavior, particularly with respect to transition to flaming. Figure 79 shows combined plots of maximum HRRs versus AHF for shredded newsprint, May tall fescue, cheat grass, and pine needles. The high maximum HRRs indicate that flaming was ignited for all of these fuels with AHFs of 40 kW/m^2 and higher. At these high AHFs the maximum HRRs were around 200 kW/m^2 for each of the fuels.

As the AHF was reduced below 40 kW/m^2 the two grasses no longer flamed, and the observed HRRs dropped precipitously to values characteristic of smoldering. The corresponding black body temperature for an AHF of 40 kW/m^2 is 643 °C. This temperature is much higher than the highest heated plate temperatures used, thus the cone calorimeter results are consistent with the fact that transition to flaming was not observed on the heated plate for these two grasses when no wind was applied. Since smoldering was present at lower temperatures for both grasses, these results suggest that flammable mixtures of pyrolysis gases for these grasses are only generated when the fuels are subjected to high heating fluxes and/or high fuel surface temperatures.

In contrast to the grass behavior, transition to flaming was observed for the shredded newsprint and pine needles for AHFs as low 10 kW/m^2. This is consistent with the observations for the heated plate experiments, which showed that these two fuels transitioned to flaming in the absence of a wind whenever glowing was observed.

The transition to flaming behaviors of May tall fescue and cheat grass were very different in the presence of a wind, with transition observed for the cheat grass and not for the May tall fescue.

The cone calorimeter results provide no indication of this difference since the observed ignition behaviors and maximum HRRs were similar for both. This suggests that while radiative heat flux measurements can provide insights into smoldering and flaming behaviors for fuels placed on a heated surface when no wind is present, they do not differentiate ignition differences that appear due to the presence of a wind.

It is significant that similar differences in transition to flaming behavior were observed for the May tall fescue (did not flame) and August tall fescue (flamed) with wind present. These two samples were collected from the same location and likely consisted of the same type of grass (there is a possibility that more than one type of grass was present and grew at different rates, depending on conditions). Even so, the results indicate that ignition characteristics for grass samples from a given location can vary with seasonal changes.

The fact that flaming is observed for the May tall fescue and cheat grass with similar AHFs in the cone calorimeter does provide a clue as to the reason for their different flaming behaviors when placed on the heated plate in the presence of wind. The observations suggest that the temperatures developed on the fuel surfaces as a result of oxidation are higher for the cheat grass than for the May tall fescue.

The cone calorimeter results for polyurethane foam were different from those observed for the cellulosic fuels. No smoldering was observed, and the material either flamed or did not ignite. For AHFs of 30 kW/m^2 or greater the ignition times were less than 50 s, increasing to several hundred seconds for AHFs between 23 kW/m^2 and 25 kW/m^2. No ignition was observed with APHs of 22 kW/m^2. This value is much higher than the minimum AHFs that ignited smoldering in the cellulosic fuels. Two distinct burning periods with different HRR behaviors were observed in the HRR time histories for the polyurethane foam that were not characteristic of the other fuels. This was attributed to the formation of a liquid component during the foam burning followed by pool burning. The higher AHFs required for ignition and the burning behavior suggest that polyurethane foam placed in contact with or near a heated surface may not ignite as easily as shredded newsprint or the natural fuels that can smolder at lower temperatures.

As discussed in Section 2, only limited information is available in the literature for comparison with the results of this investigation. The most similar studies were those of Harrison [17] and Kaminski [18], who reported limited measurements of ignition times for natural fuels when placed in contact with variable temperature heated surfaces. Harrison reported that pine needles and grass ignited in about four minutes with surface temperatures of 400 °C and 350 °C, respectively. The effect of a light wind was to reduce the ignition time. With the exception of punky wood, which ignited in several hundred seconds for a surface temperature of 270 °C, Kaminski found that most of the fuels he investigated would ignite with surface temperatures between 300 °C and 330 °C. These results are comparable to those reported here. Cheat grass was reported as requiring 120 s for ignition at 330 °C. The current measurements (see Figure 51) indicated that cheat grass did not ignite at temperatures below 380 °C when a wind was not present.

Stockstad has reported the ignition behavior of small pieces of cheat grass [20], pine needles [21], and rotten wood [19] when exposed to heated air in an oven. Cheat grass ignition was

84

observed to require 7 s for a temperature of 380 °C, while pine needles required about 15 s for temperatures as low as 300 °C. These minimum ignition temperatures are in good agreement with values for the same fuels determined in the current study, but the observed ignition times are much shorter than those measured with the heated plate.

No previous experiments were identified in the literature with which the radiative heating results could be directly compared. White and coworkers have employed the cone calorimeter to investigate piloted ignition for a variety of natural fuels including cheat grass, but measurements have been made using only one or two AHFs. [24,25,26]

Manzello et al. investigated the ignition behavior of similar fuel beds, including shredded newsprint, grass, and pine needles, in the presence of light winds due to contact with one or more glowing fire brands placed on the top surface. [22,23] Their results indicated that shredded newsprint developed smoldering, grass did not smolder or flame, and pine needles developed smoldering and transitioned to flaming only with four large glowing brands applied and with the highest wind (1 m/s) used. These results suggest that glowing fire brands placed on top of a fuel bed are less likely to ignite smoldering or flaming than when similar fuels are placed on top of a heated plate. This is the case even though heated plate temperatures were likely much lower than the surface of a glowing fire brand.

The ignition behaviors of the cellulosic fuels investigated can be classified into two broad groups: pine needles and shredded newsprint which developed smoldering and transitioned to flaming when no wind was present and the grass samples which developed smoldering, but did not transition to flaming without an imposed wind. Note that the May tall fescue/pine needle mixture had ignition behaviors intermediate between these two extremes. It is not possible to definitely identify the reasons for these differences without careful characterization of the fuels and their response to heating. However, it is possible to provide a plausible explanation. Newsprint is known to consist primarily of cellulose, while the natural fuels contain varying proportions of cellulose and lignin (see the discussion in Section 2). As discussed earlier, fuels with high percentages of cellulose are more likely to flame, while fuels with higher percentages of lignin tend to smolder. This suggests that the shredded newsprint and pine needles investigated here tend to flame because they contain higher percentages of cellulose relative to lignin than the grasses that were studied.

It is possible that the different ignition behaviors of May tall fescue (limited transition to flaming) and August tall fescue (transitioned to flaming) in the presence of a wind are due to differences in the cellulose/lignin ratios for the two grasses, with a higher ratio of lignin in the rapidly growing spring grass, or to some other type of seasonal compositional difference.

The results of this investigation have demonstrated that smoldering and flaming ignition of cellulosic fuel beds exposed to conditions chosen to be representative of those generated on OPE heated exhausts vary with surface temperature, fuel type, wind, and season. In addition to these parameters, there are a number of others that would be expected to affect ignition behavior but were not varied in this study. One of these parameters is fuel moisture. The grasses and pine needles used in this investigation had measured fuel moisture percentages ranging from 10 % to 14 %. This narrow range is consistent with the expected values for these types of dry fuels stored in a general purpose laboratory maintained near 20 °C. Such small variations are not expected to significantly affect ignition behavior. Much wider variations in fuel moisture are observed in nature, with higher values often present in living and recently cut vegetation and lower values found for dried vegetation located in higher temperature or low-humidity environments.

Babrauskas provides good reviews of the effects of moisture content on ignition behavior in general and for outdoor fuels in particular. [5] The primary effect of moisture content on cellulosic fuels is expected to be a lengthening of ignition times since a fraction of the heat applied to the fuel will be used to remove fuel moisture at the expense of heating the fuel. Once the fuel moisture has been removed, the fuel will begin to heat and eventually surface oxidation can become self sustaining.

Other important parameters not varied in the current investigation include fuel bed exposed surface area, fuel bed thickness, fuel bed porosity, spatial orientation of the heat source and the fuel bed, and the accessibility of the interior of a fuel bed to an applied wind. The potential for changes in ignition behavior due to these fuel bed properties must be kept in mind when using the findings of this investigation to assess the potential for ignition of fuels by heated surfaces.

Even though the findings of this investigation are subject to the limitations discussed above, it is possible to derive some guidance with regard to ignition behaviors expected for heated exhaust surfaces of OPE. The results indicate that ignition of natural fuels can occur for heated surface temperatures as low as 290 °C. All of the cellulosic fuels studied had lower limit ignition temperatures for some wind conditions that were within 50 °C of this value. In general, these lower limits are in good agreement with the value of 320 °C recommended by Ford as being representative of the lower ignition temperature for natural fuels. [15]

The temperature-ignition time curves observed in this study are likely to be important when considering potential scenarios involving ignition of fuels by contact with or radiation from heated surfaces. Several minutes were required for ignition (either glowing or flaming) near the lower limit temperatures for all of the fuels investigated. However, when temperatures were increased by 100 °C or AHFs by 10 kW/m^2, these times dropped to tens of seconds and became very short when heated plate temperatures or AHFs were increased further.

While clearly not representative of the worst possible cases, the ignition time curve fits to heated plate temperatures given by Equation (2) to Equation (4) should be useful for estimating times to ignition for general cellulosic porous fuel beds brought into contact with heated surfaces having a temperature range of 290 °C to 550 °C. In order to provide a safety factor, the calculated ignition times for a given surface temperature could be reduced by a factor of 2. It would also be appropriate to consider the calculated ignition times to be the periods required for flaming to develop, even though in many cases only glowing combustion would occur. Additional reductions in calculated ignition times would be prudent if winds higher than 2.5 m/s are considered likely.

The experimental findings suggest there is a correlation between the temperature dependence of ignition on the heated plate and black body temperatures corresponding to AHFs used for the

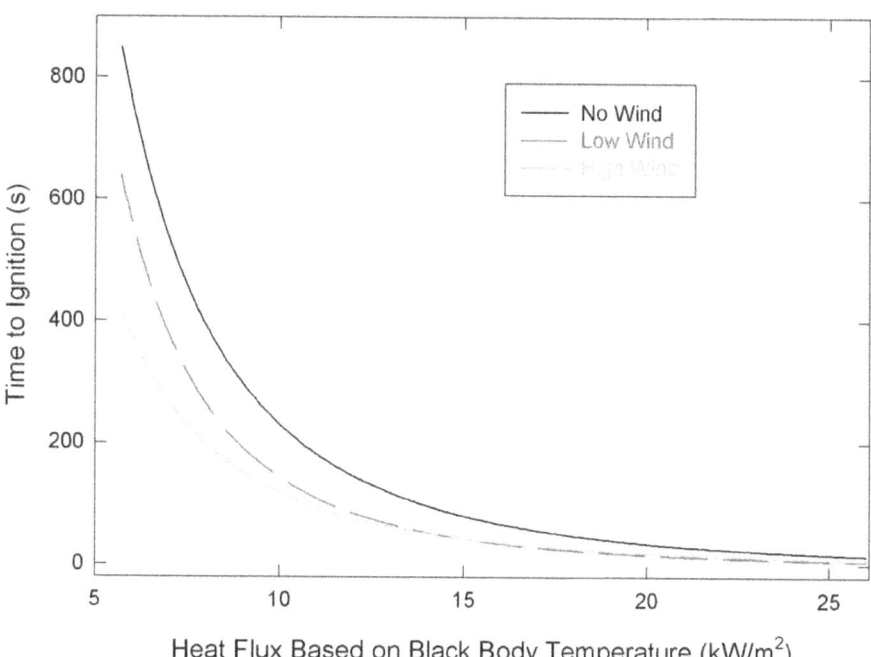

Figure 80. The curve fit times to ignition for no wind, low wind, and high wind shown in Figure 76 are plotted as a function of the AHF calculated for the radiative heat flux from a black body held at the heated plate temperature.

radiative ignition experiments. For the fuel beds and heating configurations studied, the calculated black body temperatures were slightly higher for the lowest ignition temperature and the locations of the knees in plots of ignition time versus heated plate temperature and AHF. As noted, these differences are in the direction expected based on the different fuel bed holders and directions of heat application for the two experiments.

A reasonable approach for estimating ignition times for cellulosic fuels subjected to a radiative heat flux is to use the time/temperature curves for heated surface ignition, and convert the heated plate temperatures to AHFs assuming black body thermal radiation. Figure 80 shows the results. The curve for the no wind case has a similar appearance to those determined in the radiative ignition experiments. There are no experimental data for the effects of wind on the ignition of radiatively heated fuel beds. However, given the similarities of the no wind cases, it is reasonable to expect that the effects of applying winds are captured correctly.

The determination of the ignition time/temperature and ignition time/applied heat flux relationships for cellulosic fuels was time consuming due to the large number of measurements required. It should be possible to learn a great deal concerning fuel effects with many less experiments at a single temperature. Based on the results of the experiments reported here, an appropriate plate temperatures for such tests would be 380 °C, with experiments run with and without an applied wind. Such experiments would provide relative indications for the propensity of fuels to smolder and flame in the absence of a wind along with an indication of the effects of wind on these ignition behaviors.

The experimental systems developed for this investigation have proven useful for the purpose of characterizing the ignition of cellulosic fuels. There are a number of improvements that could be made to the heated plate system. Perhaps the most important would be the use of a computer controlled system for maintaining a constant plate temperature while automatically recording the plate temperature. It would also be possible to record the currents required to maintain a constant temperature. Qualitative observations indicate that this measurement would provide insights into whether or not endothermic or exothermic processes were taking place as well as an indication for the amounts of heat generated during surface oxidation. The development of such a system should be straight forward and would alleviate the need to carefully review video tapes in order to record the temperatures.

The current findings show that HRRs from the cone calorimeter provide information concerning smoldering and flaming of these fuels when subjected to radiative heat fluxes. The time behavior of the HRR provides a good indication for when ignition occurs. Due to the configuration, it is difficult to visually observe when glowing combustion develops within the fuel bed. It would be useful to modify the cone calorimeter in a way that would allow improved detection of the onset of glowing combustion.

In order to provide higher confidence in the findings of this investigation it would be useful to make similar measurements for a wider variety of fuels. In particular it would be useful to consider a wider variety of grasses and to include various types of leaves. It would also be important to characterize the role of such parameters as fuel moisture and fuel bed density, porosity, and thickness that were not varied.

The types of studies reported here could serve as the basis for a much improved understanding of the ignition behaviors of cellulosic fuels. It should be possible to correlate fuel composition (e.g., cellulose to lignin ratio) with ignition behavior. It would be also worthwhile to examine the possibility that small scale heating experiments such as differential scanning calorimetry (DSC) and thermogravimetric analysis (TGA), which record the response of small samples to heating, correlate with the fuel bed ignition behaviors since surface reactions appear to play such large roles.

6. Acknowledgements

Many people contributed to the success of this effort. Funding was provided by The International Consortium for Fire Safety, Health, and the Environment (ICFSHE) with financial oversight by Karen Suhr. Technical oversight was provided by Petra Andersson and Margaret Simonson of SP Technical Research Institute of Sweden.

Cheat grass was harvested and forwarded to NIST by Tye Morgan working under the direction of Bob Blank, both with the United States Department of Agriculture Agricultural Research Service Laboratory in Reno, NV. Arrangements for acquisition of the fine Florida grass were made by James McNew of the Outdoor Power Equipment Institute. Sam Manzello of BFRL/NIST provided the pine needles that were tested.

The heated plate was designed and assembled at BFRL/NIST. Bill Grosshandler provided the design and Mike Selepak, Michael Smith, and Ed Hnetkovsky ordered parts, did required machining, and assembled the plate.

All experiments in the cone calorimeter were performed at BFRL/NIST by Michael Smith, who did an excellent job.

7. References

[1] Control of Emissions From Nonroad Spark-Ignition Engines, Vessels, and Equipment, Docket EPA-HQ-OAR-2004-0008, Regulations.gov.

[2] P. Andersson and M. Simonson, Scientific Evaluation of the Risk Associated with Heightened Environmental Requirements on Outdoor Equipment—Conclusions from Phase I and Proposed Plan and Budget for Phase II, SP Technical Note 2006:02, SP Swedish National Testing and Research Institute, Borås, Sweden, 2006, 43 pp.

[3] T. J. Ohlemiller, Smoldering Combustion, in SFPE Handbook of Fire Protection Engineering, 3rd Ed., P. J. DiNenno, D. Drysdale, C. L. Beyler, W. D. Walton, R. L. P. Custer, J. R. Hall, Jr,. and J. M. Watts, Eds., Society of Fire Protection Engineers, Bethesda, MD, 2002.

[4] K. M. Palmer, Smoldering combustion in dusts and fibrous materials, Combustion and Flame, **1** (2), 129-154 (1957).

[5] V. Babrauskas, Ignition Handbook, Fire Science Publishers, Issaquah, WA (2003), pp. 315-320.

[6] T. J. Ohlemiller, Smoldering Combustion Hazards of Thermal Insulation Materials. Interim Report. October 1, 1979-April 30, 1981, NBSIR 81-2350, National Institute of Standards and Technology, Gaithersburg, MD (August 1981).

[7] T. J. Ohlemiller, Cellulosic insulation material III. Effects of heat flow geometry on smolder initiation, Combustion Science and Technology, **26** (1), 89-105 (1981).

[8] D. Drysdale, An Introduction to Fire Dynamics, John Wiley and Sons, New York (1985), pp. 265-277.

[9] V. Babrauskas and W. J. Parker, Ignitability measurements with the cone calorimeter, Fire and Materials **11** (1), 21-43 (1987).

[10] V. Babrauskas, Ignition Handbook, Fire Science Publishers, Issaquah, WA (2003), pp. 833-847.

[11] Reaction to Fire Tests - Ignitability of Building Products Using a Radiant Heat Source, ISO 5657, International Organization for Standardization, Geneva, Switzerland, 1997.

[12] Standard Test Method for Heat and Visible Smoke Release Rates for Materials and Products Using an Oxygen Consumption Calorimeter, ASTM E 1354, American Society for Testing of Materials, Philadelphia, PA, 1997.

[13] Reaction-to-Fire Tests -- Heat release, Smoke Production and Mass Loss Rate -- Part 1: Heat Release Rate (Cone Calorimeter Method), ISO 5660, International Organization for Standardization, Geneva, Switzerland, 2002.

[14] Standard Method of Test for Heat and Visible Smoke Release Rates for Materials and Products Using an Oxygen Consumption Calorimeter, NFPA 271, National Fire Protection Association, Quincy, MA, 2004.

[15] R. T. Ford, Investigation of Wildfires, Maverick Publications, 1995.

[16] M. P. Plucinski, The Investigation of Factors Governing Ignition and Development of Fires in Heathland Vegetation, Doctor of Philosophy Thesis, The University of South Wales, July 2003, 347 pp.

[17] R. Harrison, Danger of Ignition of Ground Cover Fuels by Vehicle Exhaust Systems, ED&T Report 5100-15, U.S. Department of Agriculture, Forest Service Equipment Development Center, San Dimas, CA, November 1970, 39 pp.

[18] G. C. Kaminski, Ignition Time Vs Temperature for Selected Forest Fuels, Report of Work, U.S. Department of Agriculture, Forest Service Equipment Development Center, San Dimas, CA, November 1974, 6 pp.

[19] D. S. Stockstad, Spontaneous Ignition of Rotten Wood, USDA Forest Service Research Note INT-267, Intermountain Research Station, Ogden, UT, 1979, 12 pp.

[20] D. S. Stockstad, Spontaneous Ignition of Cheatgrass, USDA Forest Service Research Note INT-204, Intermountain Research Station, Ogden, UT, 1976, 12 pp.

[21] D. S. Stockstad, Spontaneous Ignition of Pine Needles, USDA Forest Service Research Note INT-194, Intermountain Research Station, Ogden, UT, 1975, 14 pp.

[22] S. L. Manzello, T. G. Cleary, J. R. Shields, and J. C. Yang, On the ignition of fuel beds by firebrands, Fire and Materials **30** (1), 77-87 (2006).

[23] S. L. Manzello, T. G. Cleary, J. R. Shields, and J. C. Yang, Ignition of mulch and grasses by firebrands in wildland-urban interface fires, International Journal of Wildland Fire **15** (3), 427-431 (2006).

[24] R. H. White, D. R. Weise, K. Mackes, and A. C. Dibble, Cone calorimeter testing of vegetation--an update, Papers presented at the Thirty-fifth International Conference on Fire Safety, Seventeenth International Conference on Thermal Insulation, Ninth International Conference on Electrical and Electronic Products, Columbus, OH, July 22 to 24, 2002 Sissonville, WV, Products Safety Corporation, 2002, 12 pages.

[25] D. R. Weise, R. H. White, F. C. Beall, and M. Etlinger, Use of the cone calorimeter to detect seasonal differences in selected combustion characteristics of ornamental vegetation, International Journal of Wildland Fire **14** (3), 321-338 (2005).

[26] R. R. Blank, R. H. White, and L. H. Ziska, Combustion properties of Bromus tectorum L.: Influence of ecotype and growth under four CO_2 concentrations, International Journal of Wildland Fire **15** (2), 227-236 (2006).